No. 722
$8.95

Amateur FM Conversion & Construction Projects

By Ken W. Sessions, Jr., K6MVH

TAB BOOKS
Blue Ridge Summit, Pa. 17214

FIRST EDITION

FIRST PRINTING—JUNE 1974

Copyright © 1974 by TAB BOOKS

Printed in the United States
of America

Reproduction or publication of the content in any manner, without express permission of the publisher, is prohibited. No liability is assumed with respect to the use of the information herein.

Hardbound Edition: International Standard Book No. 0-8306-4722-8

Paperbound Edition: International Standard Book No. 0-8306-3722-2

Library of Congress Card Number: 74-79584

Contents

A Capsule History of Amateur FM	5
ANTENNAS	
Two-Meter Collinear	8
The Frequency-Independent Beam	10
Lowering the Frequency on Commercial Omni Gain Antennas	13
Retuning Prodelin's Big Stick	14
The Power-Quadrupler Omni	17
The Sewerpipe Antenna	22
CONSTRUCTION	
Roll Your Own 2 Meter FM Handie Talkie	28
Add Two Watts to Your H/T	33
The Solid-State Carrier-Operated Relay	37
UHF Amateur Mobile Telephone	39
Quickie T-Power with Whine Filter	45
Portable Dual-Tone Digital Encoder	46
Remote Crystal Oscillator	50
Dial-On Frequency Standard	55
Controlled Charging of Ni-Cad Batteries	58
Integrated Circuit Repeater Identifier	59
Telephone Command of Repeater Operations	69
Quickie Tone Generator for Whistle-On Use	71
The Peaker-Tweaker	74
The Poor Man's Frequency Meter	76
Quickie CTS Decoder	82
Frequency Synthesis: the Modern Way to Control Frequency	85
Transistor Preamp Hi-Fi Audio from Carbon	114
EQUIPMENT CONVERSION	117
AM to FM ... in 10 Minutes!	118
Putting the NINIC Pocket Receiver on Channel	120
Split-Reed Vibrators for Simplified 6- to 12-Volt Conversions	130
Improving the Gonset G-151 FM Communicator	133

Two-Wire Remote with Zener-Stabilized Squelch	138
Converting the 450 MHz Prog Line Telephone Mobile	141
Two-Freq. and Simultaneous Monitoring GE 4ER6	143
FM on 10	147
Complete Narrowbanding of the GE Pre-Prog	157
A New Lease on Life for GE 450 Pre-Prog Receivers	164
An AC Supply for the Motorola H23 Handie-Talkie	167
Plug-In FET Circuits Replace Vacuum Tubes	169
6 Freq. Conversion: 80D & 140D Transmitters	172
4-Frequency Conversion for the 450 Pre-Prog	175
Converting the 41V	177
Converting the Handie-Talkie	182

SERVICE TIPS & INFO

The Fine Art of Receiver Alignment	188
Checking Crystal Ovens	192
Ni-Cads—How Not to Ruin Them	194
How to Get the Most from Your Mobile	200
FM Service Center	204
Defeating Desensitization In Repeaters	206
Motorola Permakays	208
Deviation Setting by Clever Estimating	209

GENERAL TECHNICAL INFORMATION

The Case for Narrowband	212
How to Know What to Buy	217
Timing Devices for Remote Control Applications	221
The Two-Way Repeater	226
The Second Input Channel	229
Touchtone Control	232
Hybrid Coupling in the Remote Telephones	234
Practical Circuit Applications Using that Strange Diode: the Varactor	239
Multiplex	245
The Inside Story of the $7 Gem	256
Touchtone—How to Use it for FM Control	263
Dials & Switches and Things Like That	268
Index	272

A CAPSULE HISTORY OF AMATEUR FM

Those individuals who joined the ranks of amateur radio within the past five years might find it difficult to believe that VHF FM was virtually unheard of as recently as 1967. And during the few years before 1967 the mode was practically the exclusive domain of a handful of experimenters who were trying to narrow that elusive gap that had traditionally existed between **ham radio** and **commercial two-way radio**.

Probably it was the gap-narrowing that kept FM from finding immediate favor among amateurs. Those who were then using FM were doing it with commercial equipment converted for use in the amateur bands. And those who were doing the equipment converting were as often as not experienced in the two-way field—they were radio servicemen, installers of commercial gear, people familiar with the ins and outs of intercom-like operation.

The traditional ham was almost the antithesis of the FM operator. To him, the style of operation on FM smacked miserably of CB. Being unaware of the technical problems that must be solved in deployment of a repeater, he had very little respect for the appliance-like manner in which repeater users operated.

The early FMer, in turn, held little respect for the dyed-in-the-wool ham, either. As far as the FM man was concerned, anyone could be a ham with very little effort; but it was the "enlightened intelligentsia" who managed to break free from the stereotyped image of a cliche-spouting, beer-drinking, horsetrading, say-nothing bundle of words who would transmit without thinking and receive without listening.

And there was just enough justification for both sides to keep the factions opposed and fed with fuel. Until 1967. In that year, a number of drastic changes began to take place in amateur radio. FM began to gain a certain respectability among amateurs—partially because more and more highly qualified technical people were beginning to get involved with FM, and partially because of the prestige accorded proponents of the mode by its monthly publication—which had grown from a mimeographed bulletin to a full-fleged slick-format journal in less than a year.

It could be that the growth of FM as an operating mode was too fast. The demand by newcomers for back issues of the monthly (FM Magazine) became excessive, and the rising number of subscribers kept the magazine's working capital from accumulating. In June 1969, Mike Van den Branden—the originator of the magazine—threw in the towel. The expenditures needed to keep the journal were more than he could bear. As Mike's partner and editor of the magazine, I gave up, too. Both of us had families to support, and since our efforts on behalf of FM radio and FM Magazine had been labors of love rather than a venture of profit, there was simply no justification for continuing.

It is conceivable that the demise of FM Magazine could have doomed FM as a mode. But Wayne Green, editor and publisher of 73 Magazine, had been watching quietly from the sidelines. He saw what few of the rest of us could see: the tremendous potential of VHF FM in terms of the growth possibilities it offered for the overall hobby of ham radio. In October 1969 Wayne appointed me editor of 73 Magazine. He was to serve as a restraint when I wanted to publish a whole magazine full of FM articles, but he did encourage careful, step-by-step gentle pushes in the general direction of FM. He did not want to alienate the readers who had not yet been exposed to FM, nor did he cherish the prospect of losing FM readers who rushed to join the ranks of 73 subscribers as a result of the magazine's very obvious flirtation with this new and controversial mode. He adeptly and skillfully walked that tightrope between **recognition of FM** and **commitment to FM**.

In doing all this, Wayne Green made some new friends and a whole passel of new enemies. But all the while FM was growing. And the other amateur journals were learning what Wayne had already learned—that FM was here to stay, and that FM opened up a whole new market for manufacturers who had been experiencing a long and crippling slump in new-equipment sales.

As FM gained popularity, back editions of the defunct FM Magazine were in demand. But of course there simply weren't enough of the back issues to go around. Not every FMer wants to buy new equipment to operate. The ceaseless requests for back issues of FM serve to attest that there are still large numbers of amateur radio operators who might like the "appliance-operator" style of communications, but who also want the fun and thrill of building and perfecting their own transmitters, receivers, antennas, tone generating and decoding circuits, and the like.

If you're one of these amateurs, I dedicate this book to you. It contains articles from the pages of FM Magazine, selected on the basis of their lasting usefulness to the builder, the experimenter, the tinkerer...you.

Ken W. Sessions, Jr. K6MVH

ANTENNAS

Two-Meter Collinear
by Bob Lans VE3BXA

In a search for a good omnidirectional antenna capable of respectable gain at low radiation angles, the Uxbridge repeater group in Canada came up with the coax-section collinear described here.

Figure 1 illustrates the general layout; Fig. 2 gives a detailed view of the connections. The cable is Amphenol low-loss 50 ohm Polyfoam (621-111 8 / U). The velocity factor of this cable is 0.8 (most others are 0.66). For this cable at 147.0 MHz the length of cable for each half-wave section (end of copper braid to end of copper braid) is 31.2 inches as per formula

$$\frac{5904V}{f}$$

For solid dielectric, the sections are 26.5 in. long.

Soldering the sections together was complicated by the low melting point of the Polyfoam. This was solved by insulating the copper braiding from the dielectric with a thin sheet of flexible fiber glass. For weatherproofing and ease of mounting, the antenna can be encased in PVC electrical conduit (1 in. diameter), with the possible addition of coating of fiber glass for strength.

Preliminary field tests were conducted using a 30-watt rig (to a 5/8-wave antenna with a low swr) and then switching to the collinear. Three reports gave a poor signal report (40-80 miles) on the 5/8-wave vertical, but quieting on the collinear, a fact that attests to the attractive "gain" advantage of this collinear antenna.

Figure 3 is a device suggested to match the antenna to the transmission line. We used seven half-wave sections and no matching device and found we had a very low swr.

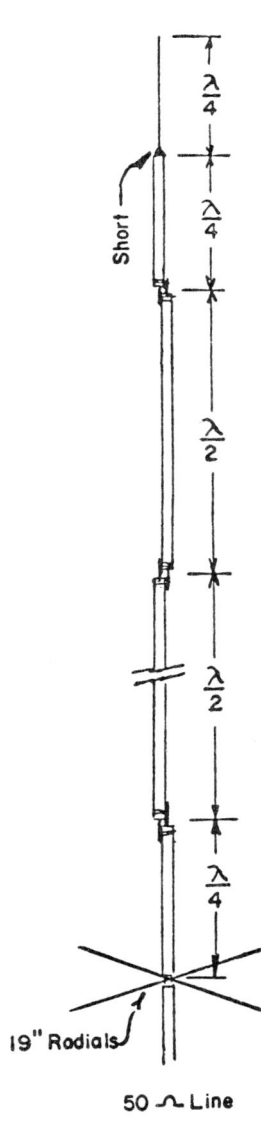

Figure 1

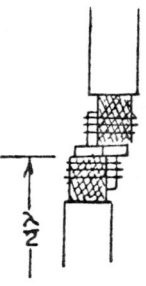

Figure 2

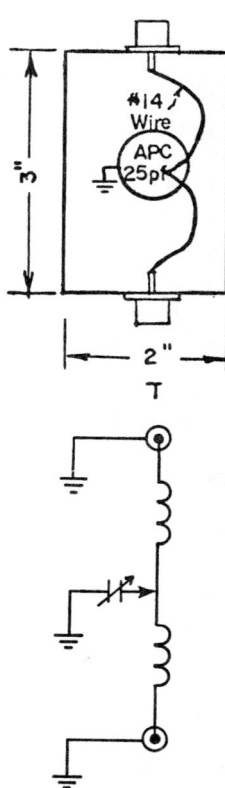

Figure 3

9

The Frequency-Independent Beam

The fellow who originated the all-band inverted vee configuration was on his way to discovering the secret of the logarithmic beam, that magic radiator so often used in today's commercial and military applications. But it was Rumsey who conceived of the "frequency-independent" concept; and his work was instrumental in reducing theory to practice by Isbell, DuHamel, and others.

There are two commonly applied methods for broadening the resonant-frequency range of an antenna. The first is the well known Q reduction technique, where the effective Q is lowered by increasing the diameter of the antenna elements. This process is a valuable spectrum widening procedure, but the ultimate bandwidth is never really unlimited.

The second method is called reactance compensation, whereby an added reactive network serves to cancel antenna reactance over an even wider range.

Both processes are used to achieve one purpose: to provide a uniform input impedance match irrespective of input frequency. So neither process, by itself, can be used to provide a constant gain and performance curve over a given frequency range. The result is merely an antenna that remains reasonably well matched, even though the gain and radiation pattern continue to vary with frequency.

Rumsey's approach was a slightly different tack to the problem, and his frequency-independent antenna designs resulted in arrays of the log periodic variety, which have the capability of providing nearly flat performance over a wide spectrum as well as a uniform input impedance.

In the strictest sense, of course, there are no antennas that are fully resonant at all frequencies. Overall antenna size governs antenna bandwidth; and, since a given frequency range depends on the size of the particular elements, it would be impractical to design an antenna with massive elements (approaching infinity) for the lower frequencies and infinitesimal elements for the very high frequencies. Thus, a frequency-independent antenna is constructed to give op-

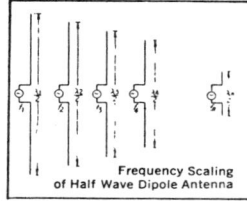 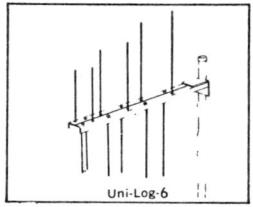

Fig. 1. Sketch (left) shows theory of log periodic beam. At right is Prodelin Inc.'s Uni-Log-6 array.

timum performance over a realistic operating range—somewhat less than the 10:1 bandwidth ratio of the traditional log periodic.

Oddly enough, there is little mysticism about the operational theory of a logarithmic beam antenna. In its simplest form, the log array is a row of ordinary dipoles, each of which is cut for a higher frequency than the preceding dipole. As with the inverted vee, the proper antenna (or dipole) responds to its own wavelength of resonance, while the other elements on the boom act as directors and reflectors. And the input impedance remains very close to 72 ohms, regardless of which dipole is excited. (Prodelin's commercial log periodic, the Uni-Log-6, has a 50-ohm balun built in as an integral part of the antenna assembly.)

The nonresonant dipoles which serve as parasitic elements (while one dipole radiates) are cooperative for several reasons. The feeding of the dipoles alternates along a two-wire line (which can also be the boom) so that those adjacent to the radiating dipole are 180 degrees out of phase with the radiator. The elements which are shorter than one-half wavelength at the operating frequency present a high capacitive reactance to the two-wire feedline and absorb only a very small amount of the energy on the antenna. Elements longer than one-half wavelength present a high inductive reactance and likewise absorb only a negligible amount of the energy. The radiating dipole, however, appears as a good impedance match to the 72 ohm line, so it absorbs the bulk of the available energy. The shorter elements then act as directors while the longer adjacent element takes on the role of reflector (whose mass is reinforced by the other longer dipoles).

In commercial arrays, the dipole lengths are arranged so there is a constant ratio between all adjacent dipoles. Similarly, the elements are spaced according to a constant determined by wavelength. While the mathematics involved in determining the ratios can become horrendous, application

Figure 2

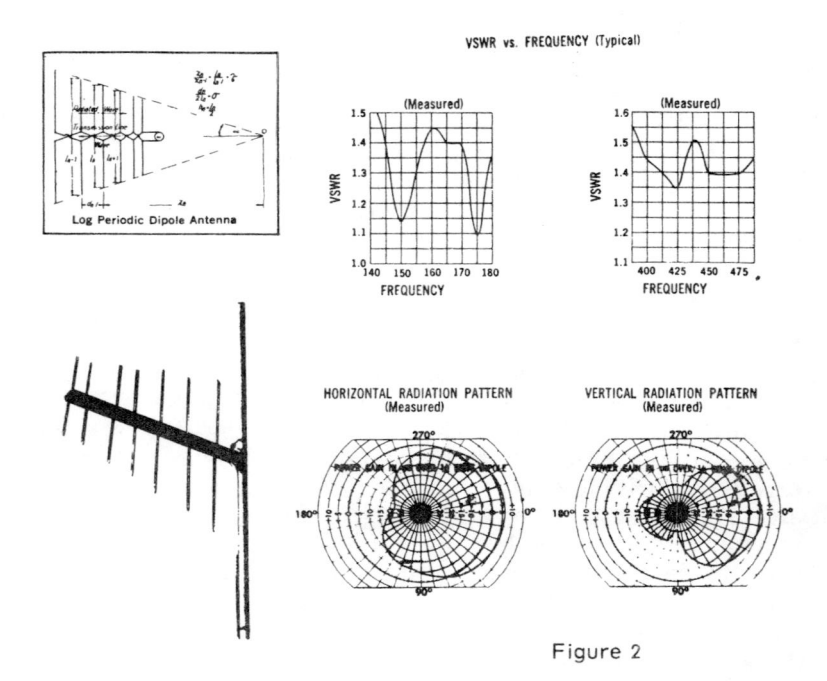

of the established ratio into a practical antenna design is no more difficult than calculating spacing and element lengths for any other type of array.

REFERENCES

1. Rumsey, V.H., "Frequency Independent Antennas," University of Illinois Antenna Laboratory, Technical Report 46.

2. Isbell, D.E., University of Illinois

3. DuHamel, R.H., Collins Radio Company

Also Cited: Prodelin Inc., Gen Catalog 644, "Antenna and Transmission Line Systems" Issue 2.

Lowering the Frequency on Commercial Omni Gain Antennas

If you've ever passed up the chance for a bargain on a good Prodelin Omni-6 or Com-Prod Stationmaster simply because it was cut for some frequency in the commercial band, you can start kicking yourself. There is a little trick you can play with capacitance to bring the frequency down to two meters without sacrificing any significant amount of gain.

The cost? Probably about a quarter! All you need is some aluminum foil plus a modest amount of electrical tape. After a half-hour with the antenna and a Bird Thru-Line wattmeter, you should have it tuned and ready for use on the frequency of your choice.

How To Do It

Cut the foil into a half-dozen or so inch-wide strips of about ten inches length. Support the antenna horizontally on two nonconducting pillars so it rests as far off the ground as possible. Connect the transmitter to the antenna through the wattmeter, and adjust the plug to read reflected power (so that arrow on wattmeter plug points to transmitter). With transmitter keyed, wrap one of the strips around the antenna at its base, and gradually shift its position for minimum reflected power. Once the position has been ascertained, secure the strip well with electrical tape.

When the first strip has been taped down, repeat the process a half-wavelength from the first. The exact half-wavelength distance can be readily determined by the sudden drop in reflected power. Secure the second strip with tape, and repeat the process all the way up the antenna.

If a 1:1 match is obtained before reaching the last half-wavelength portion of the antenna, remove all strips and start over using thinner pieces. The resonant frequency is properly lowered when all half-wavelength spots have been encircled and the wattmeter shows no power reflected.

Fig. 1. Foil strips at half-wavelength multiples over entire antenna.

Retuning Prodelin's Big Stick

by Wayne Wicks

There are many antennas available to FM'ers. One of the best for all-around performance is the Prodelin Omni-6, commonly known as the Big Stick. This antenna consists of five half-waves in phase and will offer some 6 dB omnidirectional gain. Prodelin's popular Omni-6 also offers a fantastic capture area due to the fact that the entire antenna is an active element and almost 20 feet in length (for the 150 MHz version).

I worked on one of these Big Sticks for about a week that was cut to 152.27 MHz, a local hack frequency. The trick to making this antenna work on the amateur FM channels is explained in the following material.

I know of people who have tried to retune these antennas and have come up with capacitive dummy loads. One of the secrets to retuning is not to use aluminum foil strips of the same width due to the fact that the antenna body is tapered. The foil strips capacitively load the phasing and radiating sections. Due to the tapering of the antenna, the amount of capacitance exhibited by the foil will have to be different at the base than at the top because the spacing of the elements in a capacitor along with the surface area determine the capacity of the device of which value is inversely proportional to the distance between the elements. This simply means, in reference to the antenna, the foil strips will have to be wider at the base than at the top to exhibit the same capacitance to properly tune each section.

The best way to start out is to acquire two six-foot wooden stepladders. With the antenna resting on the top steps, the antenna will rest about one wavelength above ground and will show little or no reaction with it. Another item that you will need is a Jones Micromatch, Bird Thru-Line, or other high quality bridge.

Connect the bridge at the base of the antenna and to a transmitter operating on the frequency at which you wish to resonate the antenna. Monitor the bridge in the reflected power position and take a strip of foil 2 inches in width and 6 inches long and wrap it around the base. Start sliding the foil

RADIATION PATTERN
(Isometric)

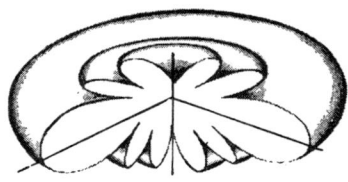

VERTICAL RADIATION PATTERN
(Measured)
(Top of Tower)

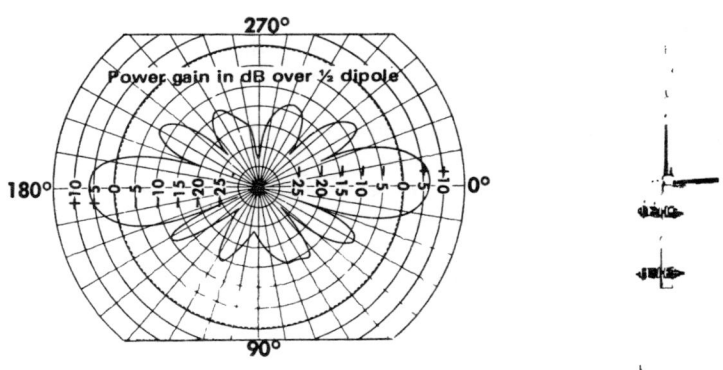

towards the top of the antenna until a dip in reflected power is noticed. Mark this spot with a piece of masking tape or the like. Continue on until the next dip is found and mark it. Continue doing this until the top end of the antenna is reached. After doing this you will have eight tape markers.

After the antenna is marked out, cut eight foil strips (to be your capacitive loads). The first strip will be 4 inches wide by 6 inches long. The next strip will be 3¾ inches wide by 6 inches long. Cut the six remaining strips, reducing the width each time by ¼ inch—the eighth strip should be 2 inches wide. All the strips should be 6 inches long. When all the foil strips are cut, apply transmitter power and monitor the vswr.

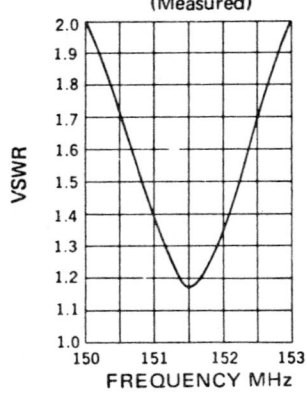

To achieve maximum efficiency in an antenna system, it is necessary to use a coaxial cable feed line exhibiting optimum performance. Prodelin highly recommends the use of Spir-O-foam aluminum sheathed coaxial cable, coupled with electrically matched Spir-O-lok connectors, to provide a "job packaged" single source responsibility for your up-to-date communication needs. Details are listed elsewhere in this catalog.

Figure 1

Starting with the 4-inch foil strip, wrap it around the antenna over the first marker starting from the base. Wrap the 3¾-inch strip around the antenna over the second marker from the base. Continue on until all the markers have been covered (the last one will be the narrowest of the foil strips).

When all the strips have been applied, the vswr should be close to 1:1. Should there still be any traces of reflected power, determine if there is too much or too little foil. Start with the first foil strip from the base and squeeze it together a little. If the reflected power decreases, continue compressing all the strips a little at a time until no reflected power is indicated.

Should more foil be required, cut eight new strips ¼ inch wider than the previous strips. Remove the old strips (but leave the markers). Reapply the new strips and recheck the vswr. When all the proper strips are on the antenna, and no reflected power is shown, start the final sealing. Use a good stretchable electrical tape to wrap the foil sections. Make sure all the foil strips are completely covered. After the tape has been put on all eight sections, spray them with a plastic lacquer to prevent moisture from penetrating.

One word of warning: When installing the antenna, make sure your support is very strong because of the high wind resistance and whipping action.

The Power-Quadrupler Omni

by Robert Shriner

The design of the antenna described herein came to light several years back; hence, I can lay no claim for its origin. Also, articles describing antennas of this type are by no means in short supply; however, improvements to the basic design, in the form of structural and mechanical modifications, make the antenna configuration worthy of new consideration.

The antenna consists basically of seven half-wave coaxial elements fed by a quarter-wave impedance matching element, and terminated with a quarter-wave section connected to a 19-inch whip. (In theory, of course, the antenna can have any odd multiple of half-wave sections, each adding gain and compressing the radiation angle.) The resulting antenna exhibits an omnidirectional, vertically polarized pattern and is extremely rugged when properly constructed, making it ideal for repeater applications.

Antennas of the type described here have been in use by the Pueblo repeater (WA0SNO) for over a year, with a coverage range that is nothing short of fabulous. The Pueblo receive antenna was constructed with 21 elements; the transmit, with seven elements. The deployment heights are 500 and 400 feet, respectively. Contacts with mobiles up to 150 miles apart are not at all uncommon.

The antenna is prepared by cutting the desired number of sections to length from a low-loss 50 Ω coaxial cable such as foam-filled RG-8/U. Cutting to exact length, and keeping all coaxial pieces uniform in length may prove a bit difficult. If the sections are prepared as shown in Figs. 1 and 2, a tubing cutter will simplify the operation.

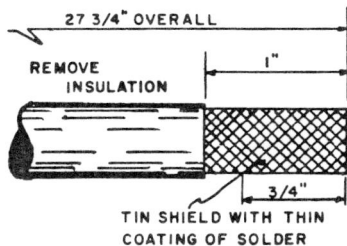

Figure 1

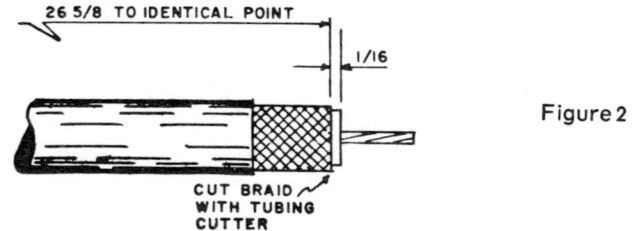

Figure 2

If the dimensions in the sketches seem squirrely, you're probably forgetting that half-wavelengths along a conductor are not the same as those of free space. The velocity factor of most popular 50 Ω coaxial types is 66 percent; thus, the individual lengths will be 66 percent of a free-space half wavelength.

Figure 3 shows a typical joint connection. I found it advisable to fashion a set of clamps to hold the elements in alignment during the soldering process. Four clothespins fastened to a flat board will do nicely. A turn or two of fine-gage copper wire around the ends of the coaxial sections will hold them in place while soldering.

To strengthen and weatherproof the joints, try wrapping a little fiber-glass cloth around the joint. Impregnate the connection with resin, and slide a piece of ¾-inch diameter by 3-inch-long heat-shrinkable tubing over the joint. Shrink the tubing carefully over the moist fiber glass; then, when the resin hardens, the joint will be fully as strong as the coax itself. With a little practice, you'll find you can make very neat-appearing and structurally sound joints every time.

Prepare the terminations of the antenna as shown in Fig. 4 (and 4a). The easiest way to do this is to take a hunk of coax about 4 feet long, strip the outer insulation off, and slide the inner dielectric and conductor out of the shield. Cut off about 15 inches of the inner conductor and flare out the wires as shown in Fig. 4a, then cut the wires so they stick out about 1∕16 inch page the insulator and slide it back into the shield. Smooth the shield down over the prepared piece and allow the

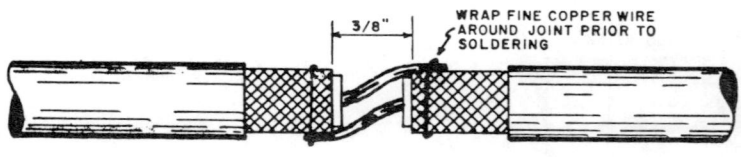

Figure 3

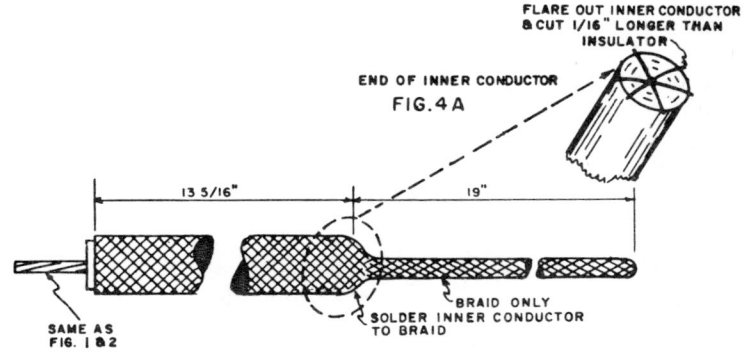

Figure 4

short wires to protrude through the shield. Solder the wires to the shield to make a short circuit at this point. Dress the balance of the shield out and finish to the dimensions shown.

Prepare the impedance matching stub as shown in Fig. 5, and file a mark in the tinned area (13 5/16 inches from the end).

Now comes the big trick! Obtain about 75 feet of 3/8-inch white braided polyethylene rope (available at marine dealers). Tape a wood dowel (1/4 inch diameter by 2 inches long) to the top end of the antenna. Start about 10 feet from one end of the rope and slip the wood dowel inside the rope. Then carefully work the dowel up inside the rope, pulling the antenna with it. Keep pushing the antenna in until the base (the point where the radials will attach) is about 1 foot inside the rope.

Prepare a brass ring as shown in Fig. 6. Equally around the perimeter of the 1½-inch I.D. ring, drill eight holes and tap for 10-32 mating.

The radials can be fashioned from brass or aluminum rod. Cut four of them to a length that is slightly in excess of a free-

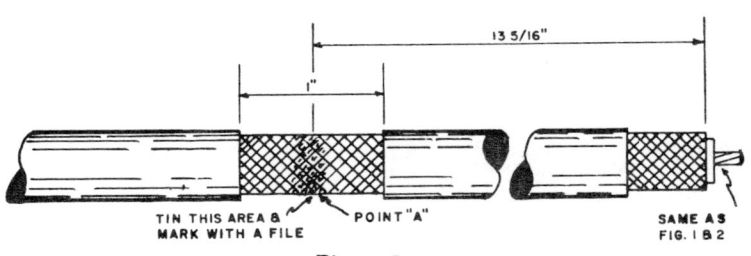

Figure 5

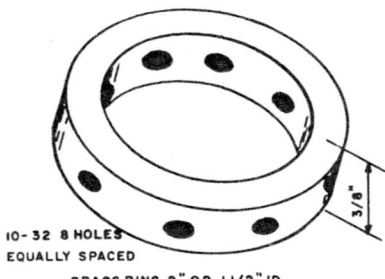

Figure 6

space quarter wavelength. Thread one end of each radial to match the threads of the brass ring, then place a locknut about an inch from the end of the threads as shown in Fig. 7. Put a 30-degree bend in each radial at the same point near the threads.

Slide the brass ring down over the rope (and antenna) to point A of Fig. 5. (This will be the terminal point shown at the base of the antenna of Fig. 8.) Insert four brass 10-32 x 1-inch machine screws in the ring and part the rope strands to allow the screws to bear against the braid of the coax exactly where it was previously marked with the file. Have a helper spread the rope so that you can reach in with a soldering gun and solder the screws to the braid. Cut off the heads of the screws flush with the brass ring and solder. Smooth the rope down over the antenna and apply a heavy coat of liquid silicone rubber between the antenna and the brass ring. This will reinforce the joint as well as weatherproof it.

Attach the four radials so that they will droop when the antenna is placed vertically. Neither the length of the radials nor the angle of droop is critical to the performance of the antenna. The values given are nominal.

The excess rope, extending from both ends of the completed antenna, will be used to suspend the creation alongside a tower. Clamp two metal braces or pipes so that they extend perpendicularly to the tower, and so that the ends are at least a quarter wavelength from the tower. Suspend the antenna between the two clamps and cut off excess rope. It is a good idea to use a heavy coil spring at the upper attach point so the

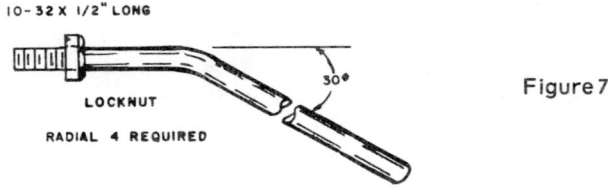

Figure 7

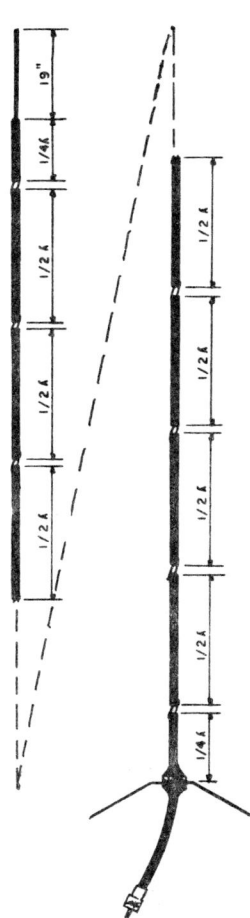

Figure 8

rope can be made taut without compromising flexibility. The spring arrangement also builds in plenty of stretch capability.

To use the antenna as a free-standing type, insert it into a fiber-glass pole. One source of supply for fiber-glass poles is a discarded commercial antenna (damaged by lightning, gashed by rough handling, etc.). These antennas, you will find, are filled with beeswax (or a similar nonhardening compound). With the application of a little heat, the wax will liquefy and pour out. To heat the wax without damaging the antenna, connect a low-voltage (1½ volts or so) transformer to each end of the old antenna conductor. As the wire temperature begins to increase, the wax exudes uniformly along the entire length of the pole.

21

The Sewerpipe Antenna
by P. J. Ferrell

The perhaps unfortunate name for this otherwise superb antenna derives from the fact that the chromium-plated brass tubing used as a matching section (normally obtained along with some funny looks from your friendly local plumbing supply house) was originally manufactured for quite another purpose.

It all starts with the J antenna, the evaluation of which is illustrated in Fig. 1. The J consists of a balanced quarter-wave matching stub feeding an unbalanced load as shown in Fig. 1c. But since balanced stubs work best with balanced loads as in Fig. 1a, some means of compensation must be provided to make the J workable.

Because of the unbalanced load on the matching section of a J antenna, the currents in the matching section are no longer equal and opposite, so the matching section radiates also. The resulting imbalance also couples rf currents to the supporting structure and the feedline, distorting the radiation pattern and making the antenna difficult to match.

The step from the J to the sewerpipe arrangement is simple. Use an unbalanced coaxial matching section for the unbalanced half-wave load. Adjusting the antenna's impedance to 50 Ω is easily accomplished as shown in Fig. 2. Chrome-plated brass pipe of 1½ inches diameter is recommended for the matching section. If ordinary brass tubing is used, then a brass plug for the bottom can be turned to fit the tubing. Mechanical details are dependent upon the materials available, and will be left to the ingenuity of the builder. Dimensions are not critical, but things should fit together tightly. The inside depth of the matching section should be about 19 inches. Keep the plastic cap (Fig. 2a) thin and use low-loss dielectric material, as this is a high voltage point. The internal feed assembly is physically similar to the gamma match used for unbalanced feed of a yagi antenna (omitting the series capacitor, of course). A clamp (Fig. 2b) completes the connection from the off-center coax to the center conductor. The height of this clamp and the center-to-center

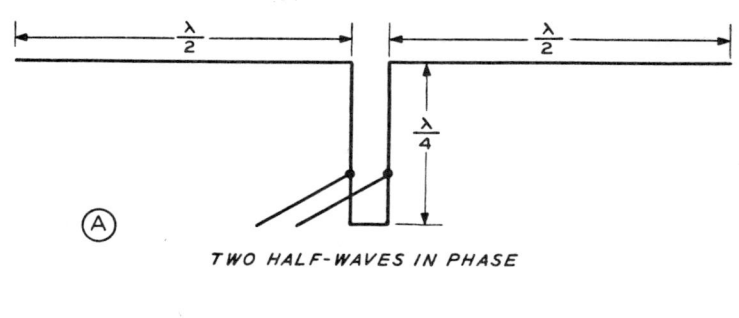

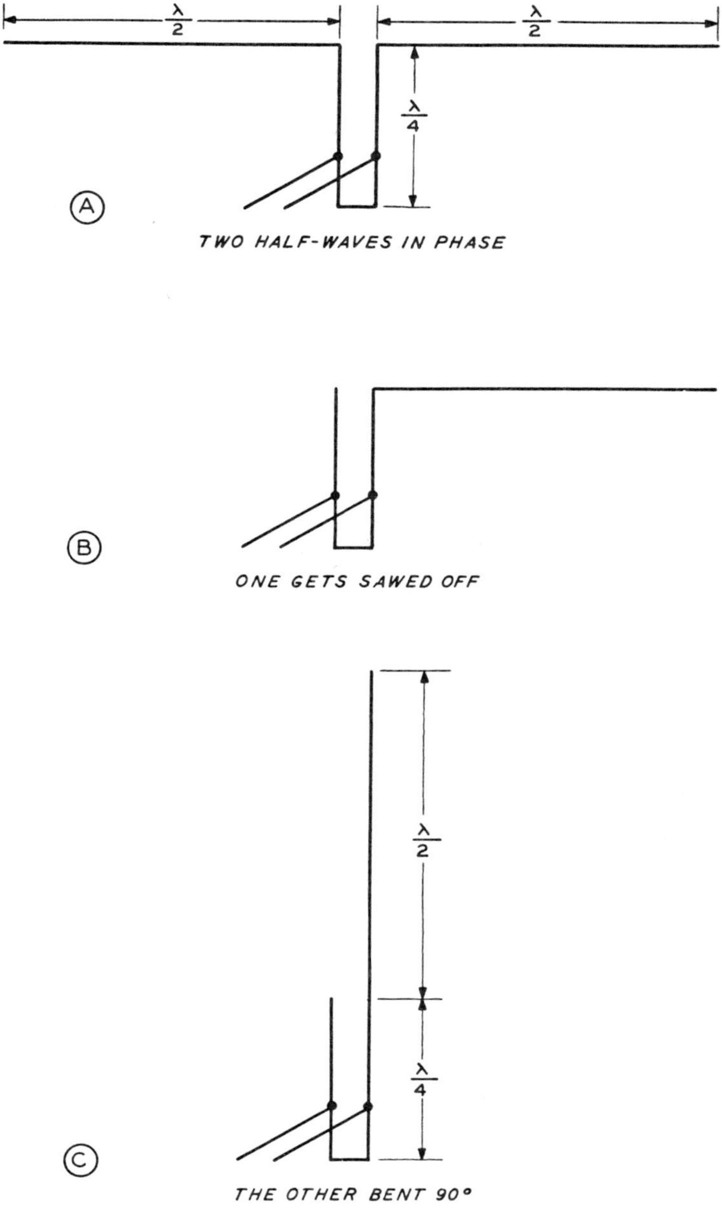

Fig. 1. Evolution of the J antenna.

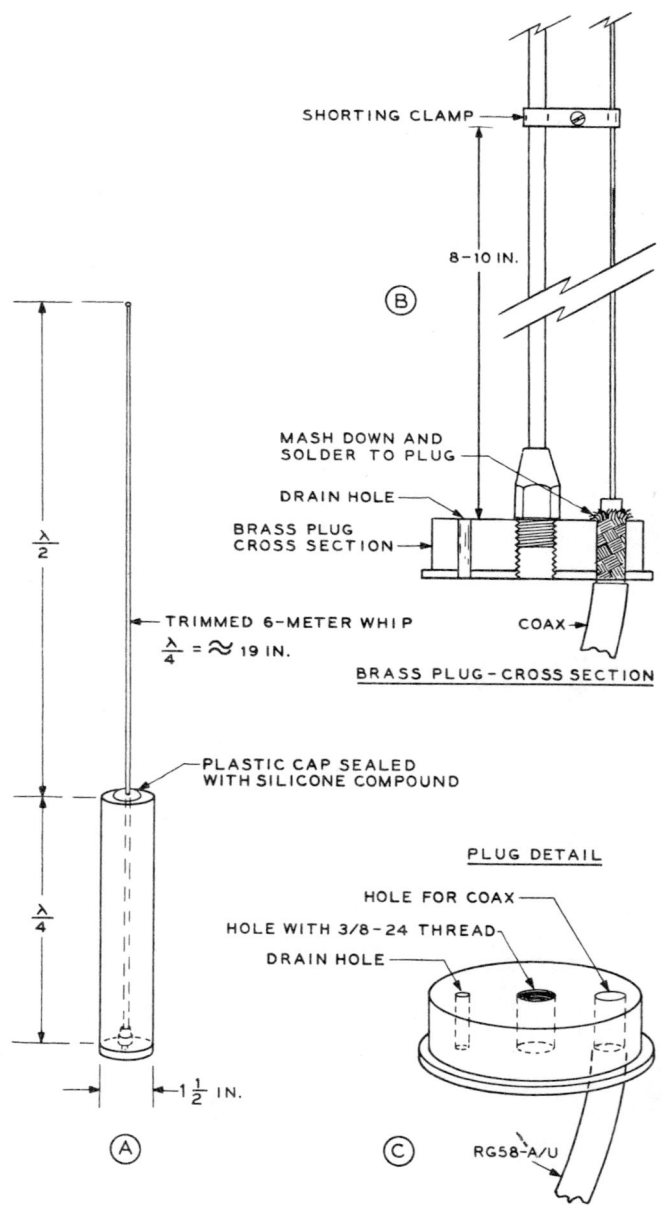

Fig. 2. The sewerpipe antenna.

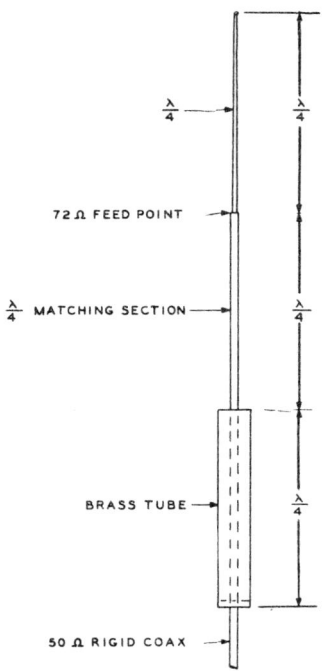

Fig. 3. Reynolds modified sewerpipe antenna.

spacing of the off-center member are varied to obtain a perfect match to 50 Ω.

An interesting variation on the original sewerpipe antenna was developed by Prof. D. K. Reynolds (K7DBA) of the University of Washington. A different feed technique is employed, as illustrated in Fig. 3; the result is an ideal antenna for base station use. Antennas of this type are in use at Byrd VLF Substation, Longwire, Antarctica, home of the 21-mile dipole. There they use 146.76 MHz VHF / FM for both on-site and station-to-station communications.

Semirigid coax is used, and the 60 Ω section is made by removing alternate half-inch sections of the dielectric inside the coax for about 16 inches. This raises the characteristic impedance from 50 to about 60 Ω. The original version is more suitable for mobile use, however, because of it's greater rigidity and mechanical strength.

Patterns taken at the University's antenna range show an almost perfect free-space dipole pattern. The measured gain over an isotropic antenna was 1.62 dB as compared with the theoretical value of 1.64 dB for an ideal dipole antenna. These antennas are unbelievably well decoupled from their supporting structure and are therefore a breeze to match. The only significant current is on the antenna itself.

Since no claim is made for extra gain, then where is the claimed improvement over other antennas? Basically, it is in the reduced angle of radiation. Practically speaking, most mobile antennas at two meters—whether quarter-, or five-eighths wavelength—have about the same gain toward the horizon; moreover, they all suffer to some extent finite groundplane effects, which act to lift the angle of maximum radiation intensity above the horizon. Thus the secret of the sewerpipe antenna's performance is its straight-out angle of radiation.

If you live in an area well covered by an accessible repeater and don't stray much, the quarter-wave whip may be just right for you. But if you need long-range capability for your mobile, give the sewerpipe a try. They have been widely used in the Pacific Northwest since 1961. The sewerpipe is ignored by all CB'ers (unlike a five-eighths-wave whip), but seems to disturb the 75-meter mobile operators for some reason Could be they think it's a new chrome-plated loading coil.

CONSTRUCTION

Roll Your Own 2 Meter FM Handie Talkie

by Dan Harger W8BCI

In constructing this transmitter, start with the oscillator and work up the chain, getting one stage going at a time until the final is reached.

The transmitter multiplies 24 times—the same as the 41V—and the same rock can be used if you have one available. The frequency of the transmitter crystal turns out to be too low at 146.94. This came as a blessing, because by inserting a temperature-compensating capacitor, Centralab TCN-30 for 30 pF, cold and hot weather drifting was greatly washed out. Most of the construction uses slug-tuned coils, which are very small and require tedious pruning to get them to the desired frequency. Grid-dipping is a problem because of loading by the transistors. The best method is to grid-dip the coil about 10 MHz above the desired frequency with the transistor out and apply power with the transistor in and see if it will tune using the grid-dipper as an absorption wavemeter.

Each multiplier should be tuned individually after the transmitter is completed.

The oscillator draws about 5 mA, the multipliers 10-15 mA each, the doubler 15 mA, the driver about 20 mA, and the final 35 mA or better. The total is about 100 mA at 10 volts.

I've given the amount of turns on the slug-tuned coils, the form size and the wire size, but each particular situation will have to be altered for your construction.

The changeover relay that was selected is the Potter & Brumfield No. 505 type RS5D which is SPDT with a coil dc resistance of 2500 ohms. It needs modification to make it function solidly at 9V, since it was designed as a 12V relay. With modification it will pull in at 7V and drop out at 6V.

As it comes, there is about 3/32 in. spacing between the armature and the pole piece (coil core) of the relay. The normally closed contact is carefully bent down until there is about 1/32 in. spacing. The spring tension is provided by a small piece of strap on the relay that doubles as the contact; t

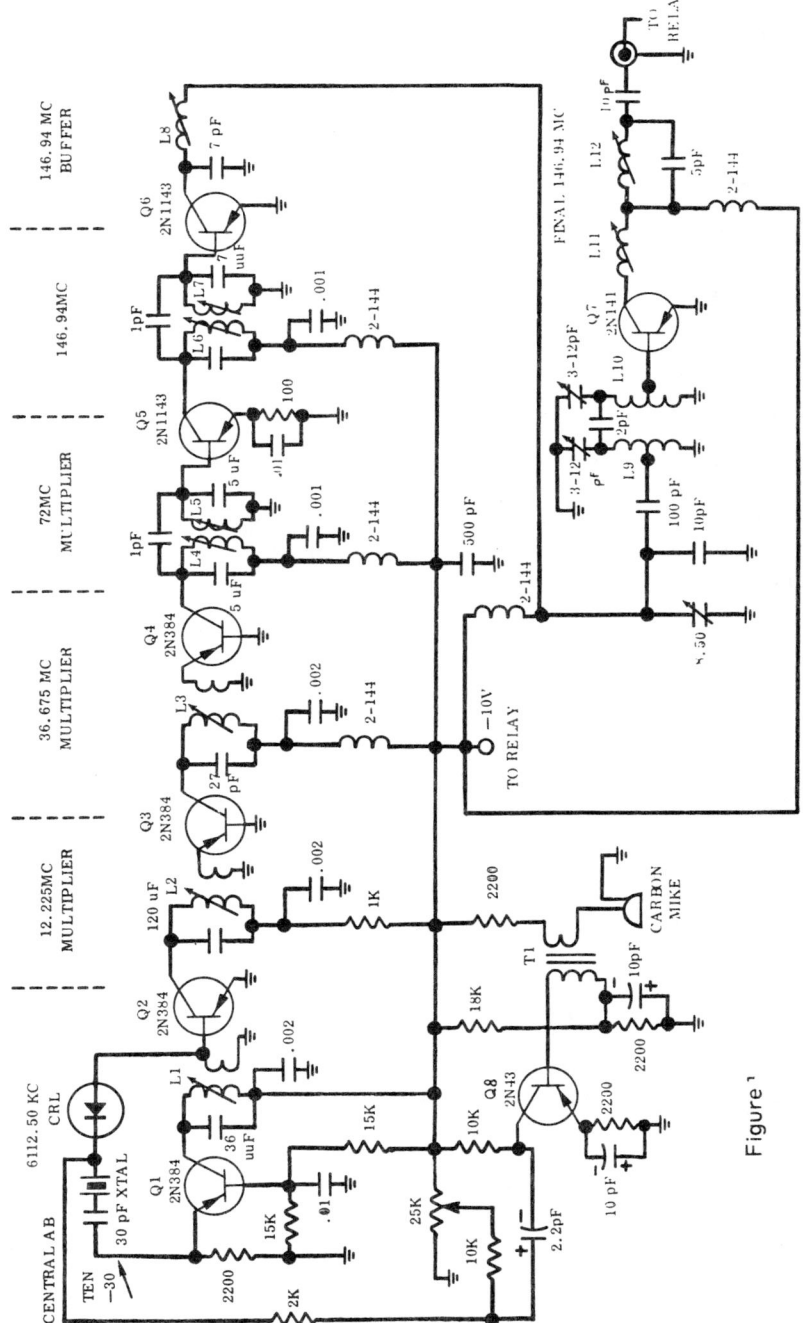

Figure 1

29

runs along the length of the armature, up and over the armature hinge, and down the back of the relay. It is secured by a small screw. This screw is loosened and the spring slightly bent so there is a little less pressure on the normally closed contact. When you bend the NC contact down, you will also have to bend the NO contact down a little.

When the relay has been modified, there should be plenty of contact pressure when the relay is unenergized and plenty of pressure on the NO contacts when the relay is energized. If you do not have enough pressure in the energized position, then you need less spring pressure. This relay has only one set of contacts, SPDT with the frame of the relay attached to the center contact. So the relay must be mounted above the frame of the chassis in the transceiver. I used the 1000 pF button mica soldered directly to the frame of the relay on one side of the capacitor and soldered the other lead directly to the center conductor of the output connector. By using a small angle bracket near the output connector as a place to solder coax grounds, the center of the receiver and transmitter coax can be soldered respectively to the NC and NO contacts through 1000 pF button micas.

The transmitter has worked very well, and the modulation is capable of at least 15 kHz deviation. Using varactor modulation, I notice that the center of the carrier does shift a little during modulation; but it is quite intelligible.

PARTS LIST FOR W8BCI TRANSMITTER:

L1—3 / 16" 25T. No. 26 enam., w / 5T link of No. 26 red slug (osc: 6112.5 kHz)

L2—3 / 16" 15T. No. 26 enam., w / 2T link of No. 26 red slug (1st mult: 12.225 MHz)

L3—¼" 10T. No. 20 enam., w / 1T link of No. 20 white slug (2nd mult: 36.675 MHz)

L4—¼" 7T. No. 18 enam., white slug (L4 & L5 3rd mult: 72 MHz)

L5—¼" 7T. No. 18 enam., white slug (see above)

L6—¼" 5T. No. 22 tinned-copper, white slug (input to buffer: 146.94 MHz)

L7—¼" 5T. No. 22 tinned-copper, white slug (see above)

L8—¼" 4T. No. 16 enam., white slug (output of buffer: 146.94 MHz)

L9—¼" I.D. air wound 5T. No. 16 tinned-copper, bare (L9 & L10 input final: 146 MHz)

L10—¼" I.D. air wound 5T. No. 16 tinned-copper, bare (see above)

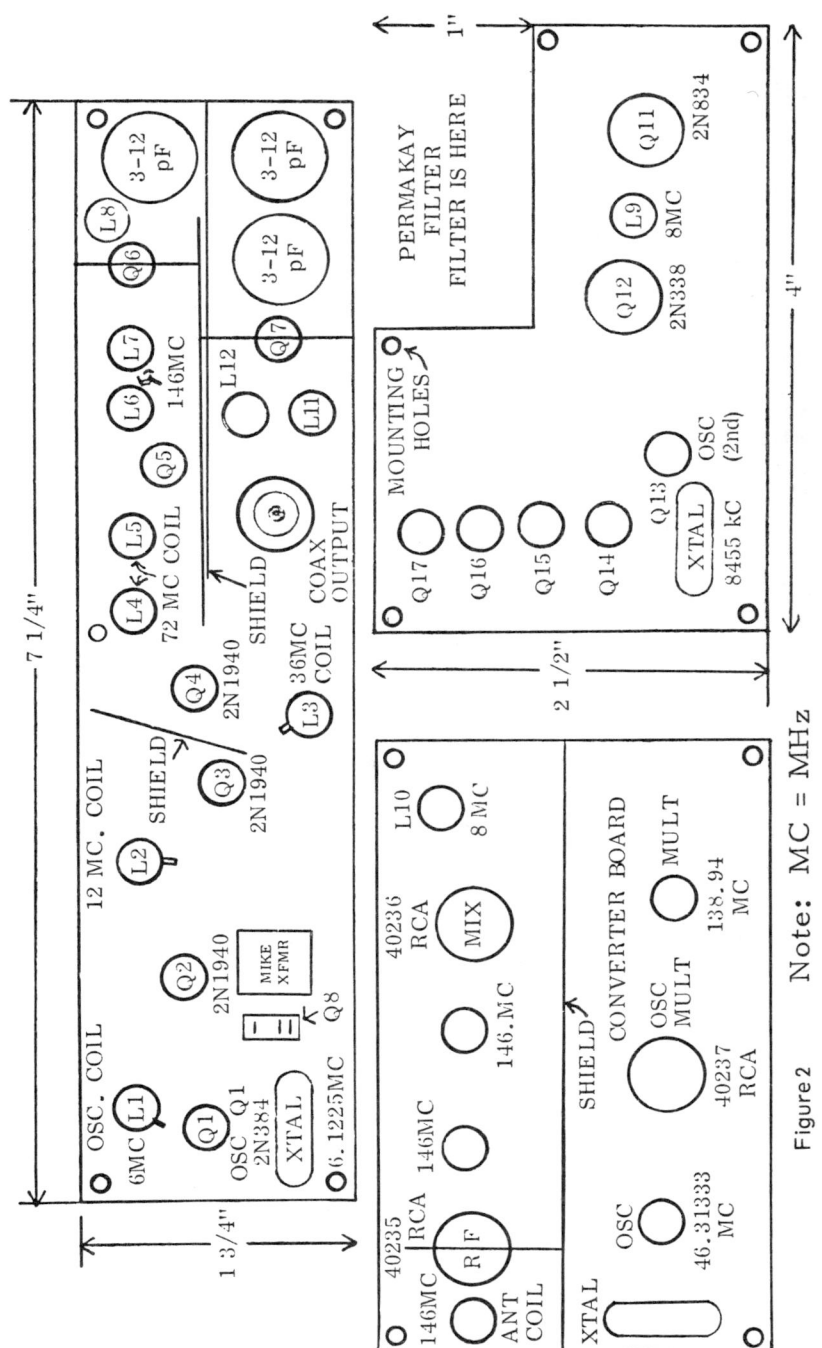

Figure 2 Note: MC = MHz

31

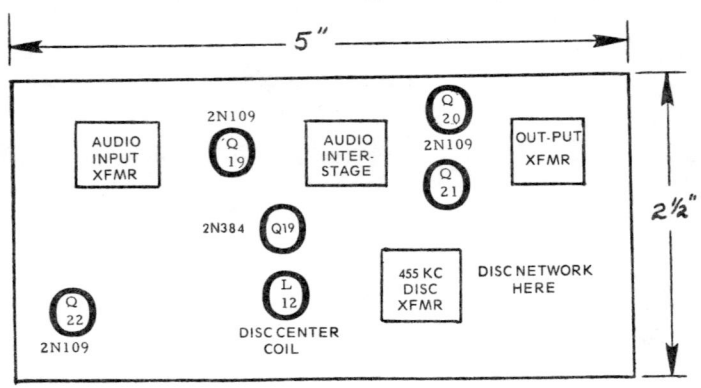

Figure 3

L11 / ¼" 4T. No. 16 enam., white slug (final output: 146.94 MHz)

L12 / ¼" 4T. No. 16 enam., white slug (antenna coil: 146.94 MHz)

The red slugs are coil forms with permeability-tuned slugs good to 30 MHz.

The white slugs are coil forms good to 300 MHz.

The variable capacitors used with L9 & L10, are ceramic, 3-12 pF.

The variable capacitor used with L8 is ceramic, 8-50 pF.

The transistor audio output transformer (T1) offers a high-impedance winding to base of Q8.

The modulator is IR type 6.8 SC20D varactor.

Add Two Watts to Your H/T

by Dan Harger W8BCI

The add-on amplifier shown here is capable of two watts output with a battery input of 10V. This is a 9 dB increase over the 250 mW of the barefoot transmitter. When the battery voltage drops a volt or so, the final output may go down to about 1.5 watts.

The transistors are RCA 40405, silicon epitaxial NPN's, each of which is capable of delivering 700 mW at frequencies as high as 500 MHz. These transistors, by the way, are exceptionally good as multipliers. The RCA 40405 is in the right price range, too. The $1.25 price tag keeps the cost of transistors for this final to only $2.50.

The transistors are manufactured in standard TO52 cases and are only about an eighth of an inch in diameter. Of course, they must be mounted to an appropriate heatsink for protection and adequate heat dissipation. A good makeshift heatsink is a small rectangular sheet of 3/16 in. aluminum panel (0.5 x 1.25 in.).

Four holes are drilled in the heatsink: two No. 14 drill-size holes for the transistors (force-fit) and two holes for passing 6-32 nylon mounting bolts. Two No. 14 holes should be placed in the center of the heatsink, and spaced about equidistant from each other and the ends of the heatsink. The mounting holes are drilled just close enough to the transistor holes so that the head of the mounting screw will hold the flange of the transistor in its hole without touching the leads. The collector and the case of the transistor are common to each other, so if the screw happens to be metal it won't matter if it is close enough to the collector lead to actually touch.

I was able to obtain some nylon 6-32 bolts and some 1/15-inch washers to space the block off the chassis. I would strongly advise against using thin mica as an insulator between the block and the chassis, as the capacitance added to other collectors might make it difficult or impossible to tune without additional modifications.

The final itself is mounted on a 3.5 x 1.25-inch copper-clad board. It fits nicely beside the batteries and leaves a reserve of space in the box.

The six variable capacitors in the final circuit allow tuning of any driver and any antenna to the final; also, it provides a means for balancing the two transistors for optimum output. (Tuned slugs could be used for this purpose if extra miniaturization is a requirement, but the variable capacitors proved the easiest way out for me.)

A shield is mounted between the input components and the output components (including the heatsink). It was possible to obtain more output with the shield than without; this was apparently due to some out-of-phase feedback without the shield. Overall stability—with the shield—is extremely good.

Construction is not really critical at all. The important thing to remember is to keep the leads short (ordinary good engineering practice) and the amplifier in a "straight line" configuration.

Checkout

When the amplifier is completed, apply 250 mW or so of drive without any voltage, using an indicating device of some kind on the output. You should be able to see drive "leakthrough"—a phenomenon common to rf transistors.

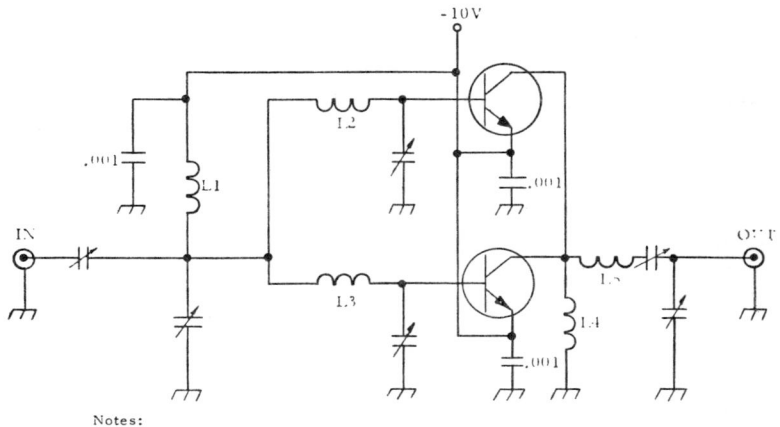

Notes:
All coils use tinned bare copper wire
All variable capacitors are 8 - 60 pF,
type not critical.

L1 - 7 turns #20, 1/8" I.D., 1/2" long
L2, L3 - 4 turns #16, 1/4" I.D., 1/4" long
L4 - 3 turns #20, 1/8" I.D., 1/8" long
L5 - 3 turns #16, 1/4" I.D., 1/4" long

Figure 1

Tune all stages for maximum output (without voltage, remember) to get the final "in the ballpark." Following this procedure, apply approximately 6V and tune for maximum output again. You should be able to produce around 700 mW. Then, with 10V on the final and 250 mW of drive, the output should increase to somewhere near a full two watts. For my own tests, I used a Bird Thruline wattmeter with a 3W plug to measure output power.

Results

The efficiency of the amplifier is quite surprising! With 2W of power from the battery (0.2A at 10V), the rf output is negligibly less than 2W. The total battery drain from the entire 250 mW transmitter and 2W final is 300 mA at 10V. To verify these figures, I have used two other wattmeters; all readings tally closely! Incidentally, I blew out the No. 49 pilot lamp used in Motorola's 250 mW load, so I installed a No. 47 in its place. The resultant vswr is not as impressive, but its life is greater—and it gives a good indication that the final is still operating. Its color is a dim orange-yellow with a 1.5W signal; the color turns more yellow as power increases.

Now for the striking results! From inside a metal structure, I could not be heard when I transmitted on the 250 mW unit. The increase to 2W, however, brought the signal up to a quieting level. Attesting to the overall communications capability is this fact: If a signal can be heard full-quieting on the Roll-Your-Own receiver, it can be worked with the 2W final. This was not always true with the straight-through 250 mW unit; the receiver has a tendency to outperform the basic transmitter.

A plus feature of the final is that it is compatible with any transmitter in the quarter-watt class. The only difficulty in such an application is that of providing a 10V power source and a means for antenna changeover. Individual ingenuity should solve both these problems with ease, however.

One last note: The amplifier oscillated when I had the wrong coils inserted. This can be avoided by applying drive without voltage (as mentioned earlier) and adjusting the variable capacitors to see if they all tune effectively. At one time, the aluminum block got too hot to touch (when the final had the wrong coils), and was oscillating with a current drain of 400 mA. Fortunately, my heatsink performed well, and the transistors lived through the ordeal. I eventually did blow the transistors, however, by shorting some of the transistor leads to ground with no drive applied. Foolishly, I was adjusting the final without cutting off the supply voltage. The proper power supply with overload protection would have prevented this.

One learns to anticipate the unexpected. A famous man once said, "Moments of despair are the lot of men with vision." Likewise, unforeseen incidents are the lot of the experimenter. But don't let them dampen your spirit. Keep trying. The results will be most rewarding!

The Solid-State Carrier-Operated Relay

Probably the most important single cause of failure in remotely operated equipment is heat. Most of us who maintain repeaters and other remotely controlled devices go to great lengths to minimize the amount of heat generated at the remote site. We replace tube-type rectifiers with silicon diodes and install carrier-operated squirrel cages to blow air over the tubes. Some of us use solid-state decoders for control of functions. But most of us still use the same old vacuum-tube carrier-operated relay to key the transmitter with each incoming signal.

Of course, it is true that replacement of a single tube with a transistor circuit won't have any great effect on the overall heat-generation characteristic of a repeater, but with each such replacement the overall "reliability figure" is improved. And replacement of a vacuum-tube carrier-operated relay (COR) with a transistor version will certainly prevent that otherwise inevitable service call for a defective COR tube.

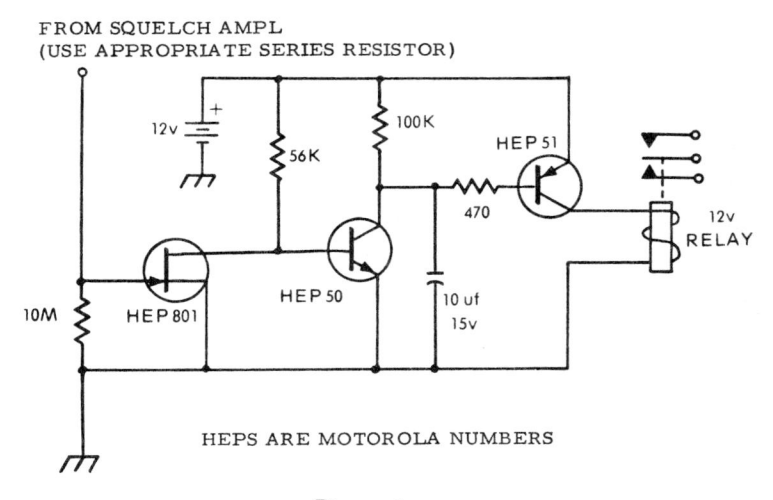

Figure 1

The advantage of transistors over tubes is sufficient in terms of reliability alone to justify their incorporation for remote applications.

Another good application for a solid-state COR is in a "walkie-talkie" repeater, a ubiquitous device which can be carried anywhere and mounted on such unlikely stanchions as tree trunks or telephone poles.

A pole-mounted transistorized repeater may sound a trifle far-fetched, but the idea does have merit. Such an approach allows installation of a low-power repeater at locations where no commercial power is available. At least one California amateur (Paul Signorelli, K6CHR) has reportedly been using a portable two-meter transistor repeater for years to successfully provide coverage in the San Fernando Valley and throughout the Los Angeles Basin. His site is high enough to provide excellent coverage, but far enough away from commercial ac lines to preclude their use as a power source.

A low-power repeater of all-transistor construction can be successfully deployed using an automobile battery as a dc source. The unit can then be kept operational by the simple expedient of exchanging batteries once a month.

Although the basic operating principles are the same for both vacuum-tube and transistor receivers, the COR shown in this circuit was designed for compatibility with a tube-type squelch amplifier, such as Motorola's conventional 12AX7 or 12AT7 variety. It can be used "as is" for most GE and Motorola receivers, but some modifications would be required to adapt it to a solid-state squelch amplifier.

The 12 volts dc required for power should prove no problem at all. A standard transmitter bias supply is an excellent source. If you have a control system, you're probably using a 12-volt dc source to drive a stepper or operate transistor circuits, anyway. If not, a small external battery will do the job nicely.

There are no critical components in the solid-state COR. The mechanical relay can be any low-current device that will pull in with less than 12 volts applied, so long as the coil resistance is not greater than 500 ohms or so.

UHF Amateur Mobile Telephone

Remote operation is getting a lot of attention these days, and an unprecedented number of amateurs are being won over to this exciting new phase of our hobby. A great many construction articles have been published over the past several years which include compatible automatic telephone interconnections. But all these articles were written with the thought that the telephone portion of the system was an add-on to a more complex setup. From the letters and phone calls I've received, I have come to the conclusion that most remoters are interested in the automatic patching feature as a primary function rather than a secondary one.

This fact has posed several problems because, for a single-purpose remote telephone, the circuits I've published in the past have incorporated extra relays and an unnecessary stepper switch. Finally, after a ham called me several times from Harrisburg, Pennsylvania to get additional circuit information, I decided to redesign the remote control circuitry for exclusive telephone use.

In the previous articles there has been a deficit of information surrounding the key elements of control: encoders and decoders. This lack of data was no accident; information just hadn't been available.

One industrious amateur in the Pasadena area (Bob Mueller, K6ASK) decided to do something about it, and proceeded to design his own encoder and decoder. With a lot of ingenuity, patience, skill, trial and error, and a few of the other necessary commodities for inventiveness, he finally came up with some acceptable circuits, but his decoder was still far from perfect for telephone use.

Lee Coltin (K6VBT) adapted Bob's circuits for use in his own remote system. Since that time, Bob and Lee have had the opportunity to refine the decoder by improving it's selectivity and frequency stability. They also incorporated a few significant improvements in their tone encoder.

With a good encoder and decoder, and a specially designed automatic phone patch circuit, the time seemed right for an all-encompassing article on a remotely controllable telephone system. So this is it.

SYSTEM REQUIREMENTS

There is a hard set of FCC-imposed constraints that usually determine the manner in which a remote system will take shape: Your system will probably include a UHF repeater on a hilltop or in a tall building accessed by a few mobiles and a control base station. It is unlikely that you will use two meters to access your home telephone. Here's why: The primary requirement for a remote system of any type is that control be accomplished from 220 MHz or above. Since gear is hard to come by for 220 MHz, most amateurs use 450 MHz (where gear is both readily available and inexpensive).

Even if you could rationalize the legality of two meters, it would be unwise to use it for phone patching. There is little privacy there, so your phone's usefulness would be seriously inhibited in this regard. Also, the two-meter region is heavily populated, so you would be subject to interference—accidental and otherwise—as well as possible phone use by unauthorized stations.

A phone system does not lend itself well to incorporation in a two-meter repeater system, either. What happens to the repeater when someone wants to use the telephone? What happens to the repeater when phone calls start pouring in? You can't just tie up the repeater so one man can make exclusive use of the telephone; it wouldn't be practical. The simple truth is that telephones belong where the activity is minimum and where control is sanctioned.

Use of 450 MHz will probably rule out installations of the control receiver and phone-access circuitry at your home; the range just wouldn't likely warrant the effort. The UHF bands are strictly line-of-sight; thus, if your system is to be successful and useful, you'll get a spot on the top of a local hill or in the highest building in town. There, you'll have your telephone, remote control repeater, and your decoder / access circuits installed.

There is no real need to go into the construction of the hilltop UHF repeater, because the subject has been covered with reasonable thoroughness previously.

Your mobile system should be no problem, either. It should be set up for duplex operation if you want a sophisticated remote telephone system. A good system is one where the party on the other end of your phone conversation can't tell you're using a phone patch. This isn't too difficult if your access stations are all capable of full-duplex operation.

The one area of control and use that does seem to have been neglected is the dial hookup. A telephone dial normally has at least two complete sets of contacts. One set is for

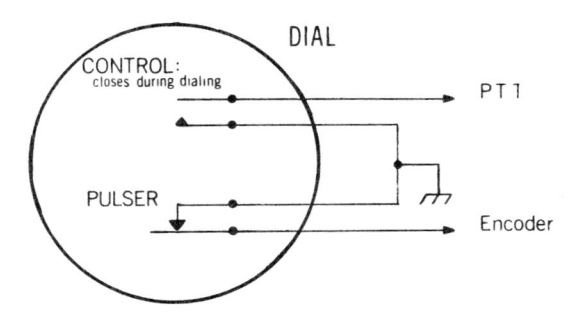

Fig. 1. Dial.

pulsing the tone encoder; the other is used to key the push-to-talk. Figure 1 illustrates this connection scheme.

When you examine Fig. 1, one thing should become immediately apparent: The pulser contacts are normally closed and open only during the pulsing process. Since the encoder must generate tones only during pulsing, the dial pulser may at first seem incompatible with a remote control system. It isn't. Take a look at the encoder circuit shown in Fig. 2. The pulser is used to keep the encoder output at ground potential until the contacts are opened. The 100K resistor between this connection and the phase modulator grid (or mike amplifier) keeps the grounded line from affecting transmitter audio level.

The tone unit itself is actually on all the time. (That is, the unit is always on when the receiver is on because the drive

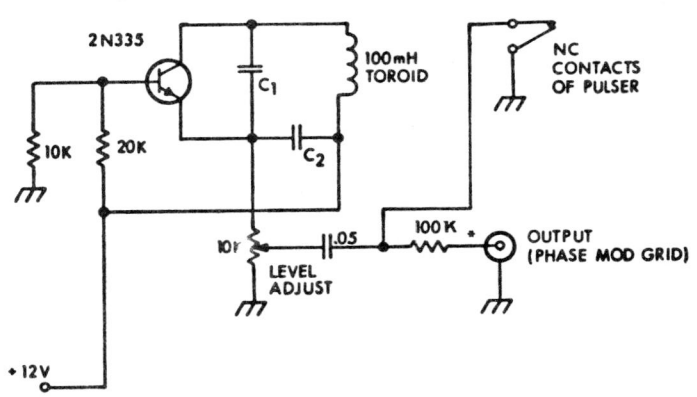

NOTE: C_1 & C_2 .02 TO 1.0 DEPENDING UPON FREQUENCY REQUIRED

*SHORT LEAD BETWEEN RESISTOR & OUTPUT

Fig. 2. Single-tone oscillator.

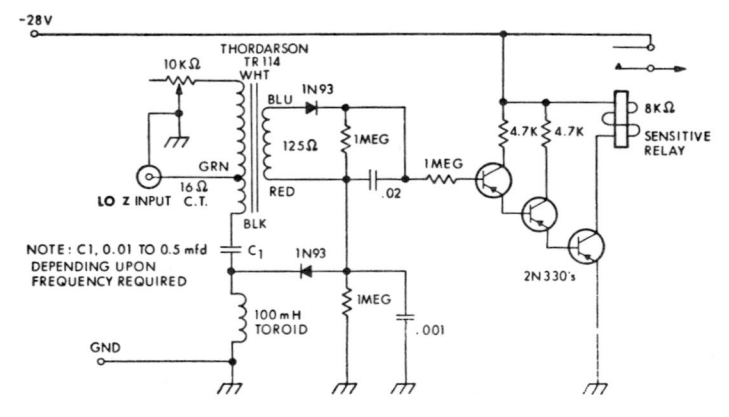

Fig. 3. Single-tone decoder unit.

voltage for the encoder is presumably obtained from the receiver or transmitter filament supply.)

An important point to remember when connecting the push-to-talk to the dial contacts is to use the set that "makes" first and "breaks" last; otherwise, you may find the transmitter won't stay on the air long enough to allow the complete pulse train to be transmitted.

ENCODER

The encoder shown in Fig. 2 is K6ASK's improved version. It has the features of high Q for stability, a single transistor for simplicity, and tiny components for miniaturization. The circuit can be easily constructed on a circuit board or perforated phenolic sheet. The 10K potentiometer in the output allows the gain to be set to whatever level is right for your own system. It can be eliminated at the sacrifice of possible incompatibility with your existing deviation level setting.

DECODER

The function of the decoder is simply to provide a relay closure with each incoming tone signal. If a "9" is dialed at the control station, the decoder responds by closing a relay nine times at the same rate as the dialed tone pulses. The most important parameters of a good decoder are frequency stability and signal selectivity. The decoder shown in Fig. 3 does have these desirable characteristics. This is what K6ASK refers to as his Mod II unit. It has a variable gain control on the input for establishing the proper sensitivity of the decoder, and a dc amplifier on the output for providing solid drive to the pulser relay.

The decoder is designed for a 28V power source. When properly driven, the unit will respond to pulses within a bandwidth of about 50 Hz (depending on sensitivity setting), and will reject all other tone signals.

Bob designed his decoder to accept low-impedance tone signals directly from the speaker terminals of the remotely situated receiver. Signals of the proper frequency and level are passed through the frequency-sensitive circuitry to the dc amplifier to key the 8K plate relay. This current-operated relay provides the pulses for controlling the automatic telephone.

AUTOPATCH

Figure 4 shows the schematic for the automatic patch. Here's how it works: When the telephone rings, the ac ring voltage is rectified (through isolation capacitors) to drive a sensitive plate relay (K1) that triggers the transmitter push-to-talk circuit as well as a special "ring" oscillator.

The plate relay keys the push-to-talk through a diode so the "ring" oscillator won't be triggered when the transmitter is keyed through the normal carrier-operated function.

To place a call or respond to a phone ring, the operator transmits a continuous tone for 0.5 second. This is achieved with the phone dial at the control point by bringing the digit "1" to the finger stop—which keys the mobile transmitter and causes the hilltop COR to turn on the repeater transmitter—and turning the dial counterclockwise just far enough to open the dial's normally closed contacts. While the contacts are

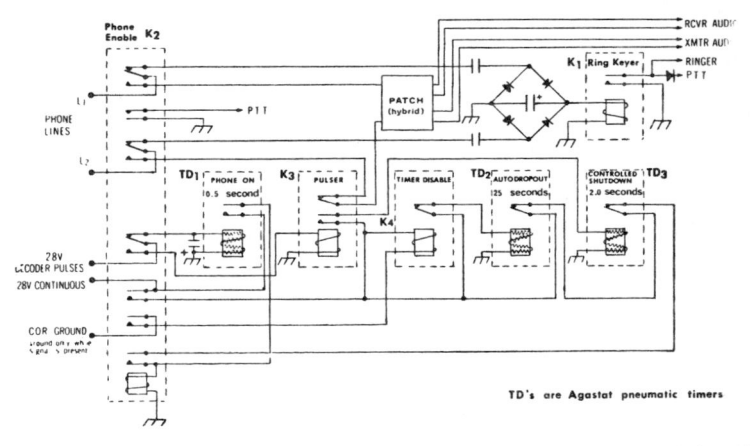

Fig. 4. Automatic telephone-operating portion of remote phone patch with fail-safe features.

43

open, a continuous tone will be generated. At the end of the half-second period, the phone will be engaged by the timer (TD1) and the dial can be released.

When the phone comes on, the telephone enable relay (K2) is energized and latched; the phone lines are disconnected from the rectifier and fed directly into the phone patch. The decoder relay is coupled to the phone pulser relay (K3) so that additional dialing will pulse the phone line exactly as a local dial would.

As long as the user transmits a signal the telephone will stay on. Presence of his signal provides a ground signal to the timer disable relay (K4) to keep the phone patch engaged. If he drops carrier, the timer disable relay is released and the shutdown timer (TD2) is started. The shutdown timer allows the phone patch to stay on for 25 seconds without the presence of a carrier. If no carrier appears by then the timer pulls in and shuts the whole system off.

The last timer in the circuit (TD3) allows the phone to be shut down by operator command. A 1.5-second tone pulls the timer in and interrupts the latching voltage on the telephone enable relay. Thus, the system includes a positive shutdown capability.

When selecting the phone patch to incorporate into your remote system, be sure to use a hybrid type. Hybrid patches are important because they allow a very definite and positive audio null and prevent feedback.

There are many manufactured hybrid types available and a great number of circuits for their construction. From my own experience, I strongly recommend the Waters hybrid patch. Some of the others I have tried offer minor impedance mismatches which result in significant audio problems during phone use. If you're sure you have good audio transformers, go ahead and build your own. If you are not sure, you'll probably be better off buying the Waters unit. It has a built-in compressor-amplifier.

Quickie T-Power
with Whine Filter

A while back a friend loaned me a circuit for an "easy" receiver T-power supply. It looked so simple, in fact, that I merely filed it. Last week I found myself in need of a mobile receiver B+ supply, so I dug the circuit out of the file. I put it together, with junkbox parts, and was pleased to find that it REALLY DOES work.

TI can be any old filament transformer with a 12-volt centertapped winding capable of several amps. This is the good part, as transistor-type transformers are usually quite expensive ($12-$20). Be sure to use a good heatsink for the transistors, as they will dissipate quite a bit of heat.

Values for the other circuit components are not at all critical, and should be chosen based on your available supply of parts.

The rf choke in series with the B+ lead on the output of the supply is an important part of the circuit, and can be used on any type of T-supply to minimize transformer whine. The switching frequency of the filament transformer will be somewhat less than that of a conventional specially designed transistor transformer. This factor coupled with the fact that the transformer is probably not potted, will add to the audiofrequency noise produced by the unit. The rf choke is very effective in swamping out the whine, and the result is a virtually noise-free dc voltage.

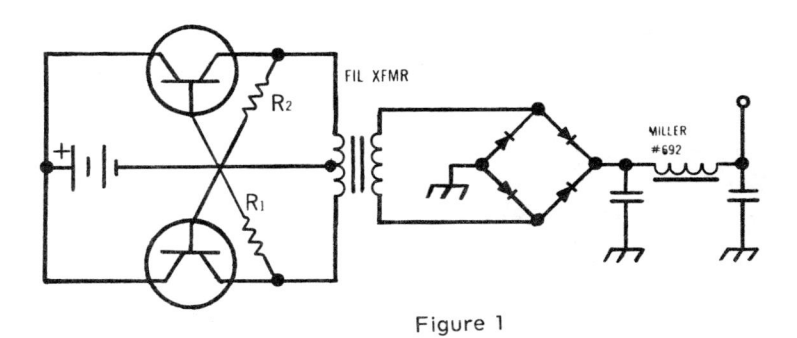

Figure 1

Portable Dual-Tone Digital Encoder

by Gary Hendrikson W3DTN

Many remotely controlled systems around the country use Secode equipment for performing the various selection functions. However, it is difficult to find a large enough supply of encoders to keep up with the increasing demand. This article shows two simple solutions.

The hand-held digital encoder shown in Fig. 1 is small enough to be used in any control application—at the fixed station, in a mobile, or with a walkie-talkie; yet, it is simple to build and is highly reliable. The unit is unique in that its tone output is audible; the sender can be placed near any control-transmitter microphone to effect the selection of remote functions.

The "speaker" on the portable sender is the receiver element from a conventional "500 series" telephone handset. I found this particular earpiece to produce the greatest output for a given input level. It will easily drive a microphone to full modulation.

The Secode system for which this encoder was designed employs the two-tone concept: as the dial is moved to the finger stop, the first tone is generated. This initial tone is decoded at the remote site to pull in the ratchet relay. The second tone is emitted as a pulse train when the dial is

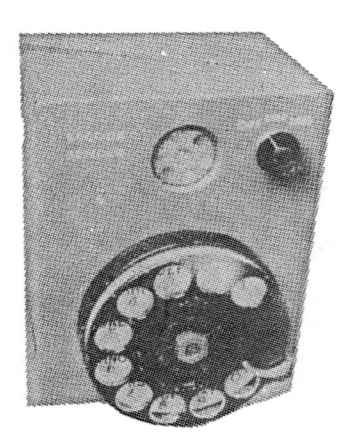

Fig. 1. Hand-held Secode-type sending unit uses telephone earpiece for audible tone output.

released. This tone causes the rotary wheel at the decoder to step to a position corresponding with the number of pulses transmitted.

As can be seen in Fig. 2, the unit can be built into a considerably smaller chassis then the one I used. The size, for practical purposes, is limited only by the dial, the earphone, and the battery requirements. Miniature dials are available on the surplus market, but often these are not provided with enough contacts to perform the dual functions of dialing and keying.

Figure 3 shows the schematic for the portable sending unit. The single transistor is not critical; it can be any PNP type with an h_{FE} of 50 to 100. An NPN transistor can be used if all supply voltages, diodes, and other polarized components are reversed.

The 88 mH toroid is the standard telephone loading coil familiar in RTTY circles. It consists of two windings; the adjacent ends at the center are tied together to obtain the single centertap.

The frequency-determining capacitors should be good-quality Mylar types for best temperature stability. Small values can be paralleled here as necessary to produce the exact frequency of oscillation. One capacitor and the switch itself can be omitted for single-frequency encoders. The two tone frequencies listed are standard with Secode.

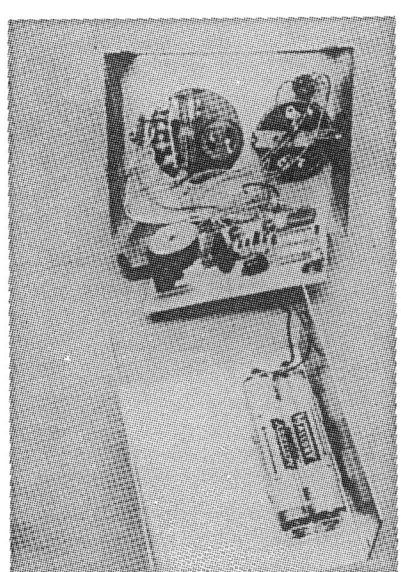

Fig. 2. Photo shows construction of tone unit.

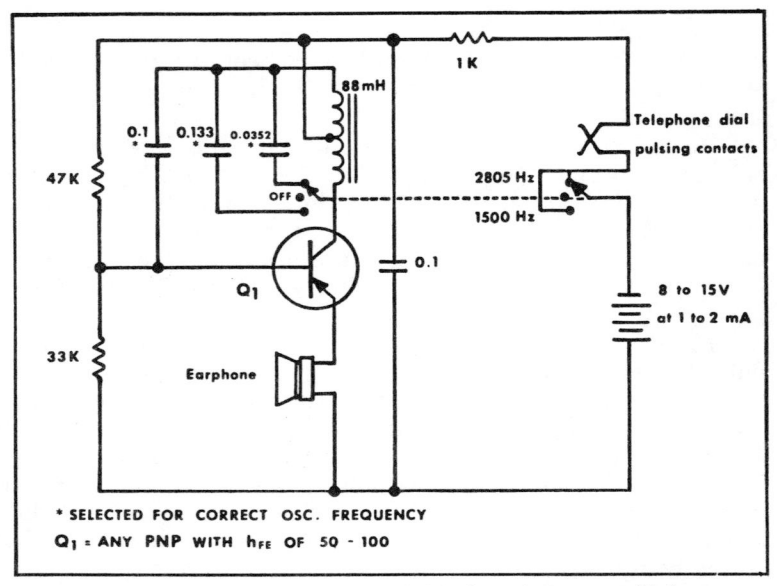

Fig. 3. Circuit diagram for portable Secode-type oscillator.

Figure 4 shows a more complex configuration that is ideal for mobile use. In the mobile version, relays K1 and K2 can be a single unit if you have a relay with sufficient contacts to perform the required switching functions.

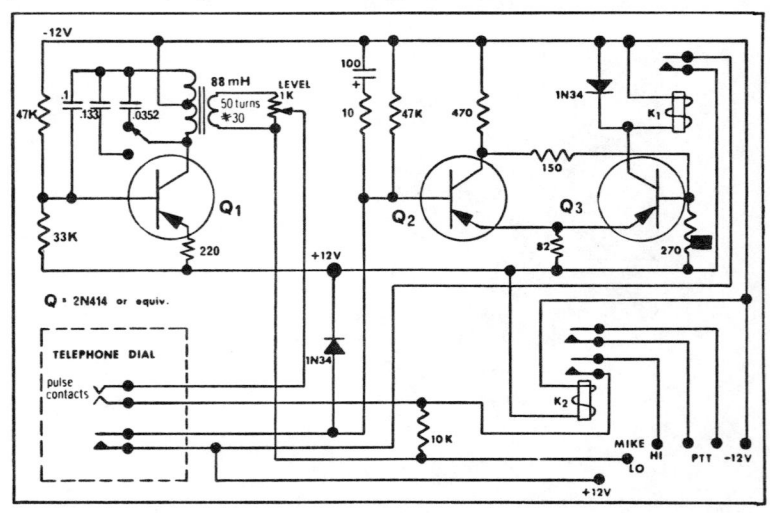

Fig. 4. Schematic diagram for mobile Secode-type encoder unit. (This version contains a Schmitt trigger device for timing transmitter-on periods.)

The two-transistor Schmitt trigger is an automatic-shutoff timer which can be set for just the time required to perform a dialing sequence. The purpose of this is to key the transmitter as soon as the dial is moved and to hold it on the air until the full dialing sequence is completed. In the hand-held version, this function is performed manually with the on-off switch.

The 100 uF capacitor in the mobile sender can be changed to effect variation of the period if the shutoff time is not to your liking. Two seconds is usually adequate, however. The transistors used for the Schmitt trigger can be the same type as that of the oscillator itself.

Layout is not critical, and construction can be by any convenient means, either with perforated phenolic board as I used, printed circuit, or solder terminal. Regardless of how you build it, though, you'll find it an effective means of joining the "squawk-teek-teek-teek" group.

Remote Crystal Oscillator

by Sam Craig W2ACM

If your VHF/FM gear is of Pre-Progress vintage and you have been wondering how to add multifrequency capability, here is your answer. This article describes a remote crystal oscillator with switch selection of up to four crystals. Since the original unit was intended for the transmitter of an ES-1B two-meter FM unit, 3 MHz crystals were used.

The circuit should work equally well with crystals of other frequencies (up to 15 MHz or so,) and it can be applied to receiver local oscillators as well as transmitters. It is not designed to work with overtone crystals, however.

The circuit consists of a transistorized Pierce oscillator coupled to an emitter follower for isolation and low output impedance. (See Fig. 1.)

The oscillator unit may be mounted in any convenient location and coupled to the transmitter by a length of ordinary shielded cable. Neither the type of cable nor its length appears

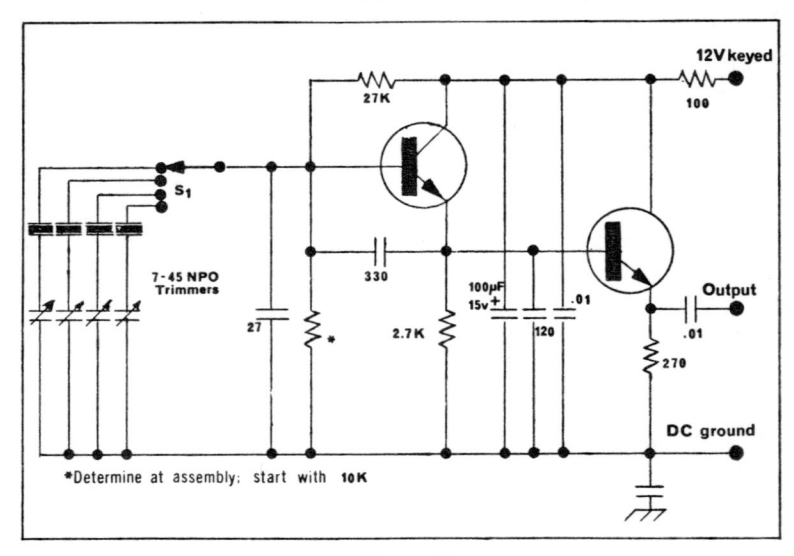

Figure 1

to be critical. In the original installation, a four-conductor shielded cable having a length of about six feet was used. The output impedance of the circuit is low enough to tolerate the shunt capacitance of 15 feet of shielded cable or coax without any drastic reduction in output voltage. Impedance matching is not necessary since optimum power is not a requirement.

The rf output from the remote oscillator is fed directly into the grid of the former oscillator tube in the transmitter, so that this stage now becomes a buffer. The cathode of this tube (and screen, if a pentode) should be bypassed to ground for rf if not already done in the existing circuit. In my installation, the grid current in this stage was not as high with the remote oscillator as it was originally, but the drive level in the subsequent stages was unaffected.

The output frequency is not affected by the load, and changes only a few hertz when the supply is varied between 5 and 15V. Although the output voltage increases as the supply voltage is increased, the use of more than 15V is not recommended.

Figures 2, 3, and 4 show the complete plans for the oscillator. Once the etched circuit boards are made, assembly of the unit is quite simple. After soldering the components to the main circuit board, attach the output, power, and ground leads and mount the board in the minibox. Be sure that leads and solder do not protrude more than 1/16 inch on the foil side of the board. Then attach and wire the connector (J1). Next, the rotary switch (S1) should be put in place and connected to the main board.

The remaining two circuit boards with the crystal sockets and trimmers are fastened together and mounted last in the box. The front of this assembly is supported by the leads from the boards to the switch, so 20 or 18 AWG solid wire should be

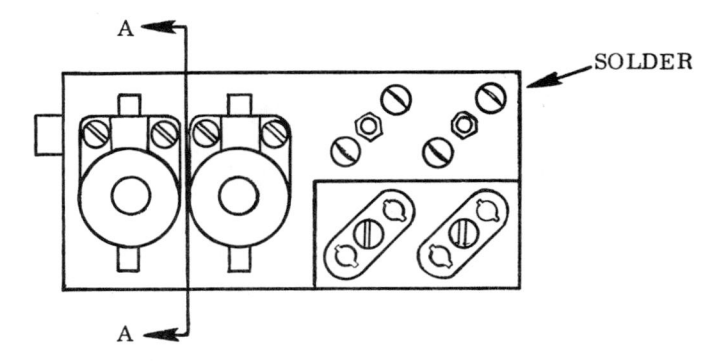

Fig. 2. Mount trimmers on foil side, crystal sockets on opposite side.

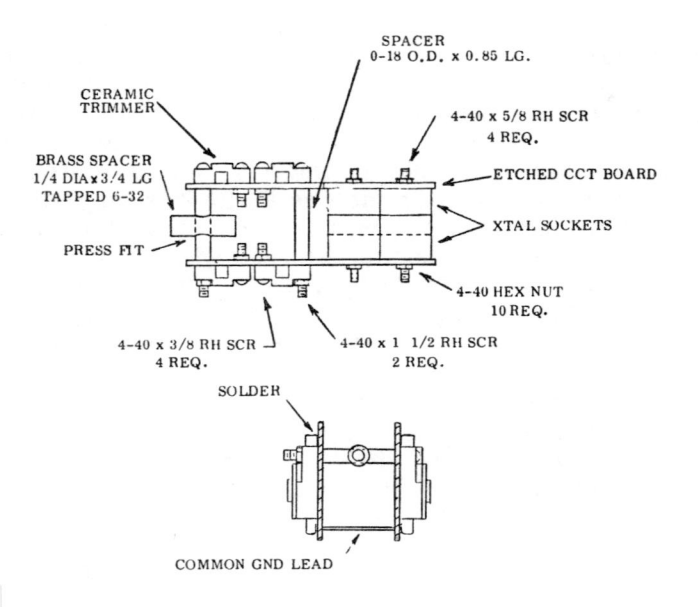

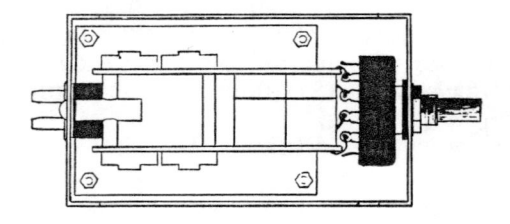

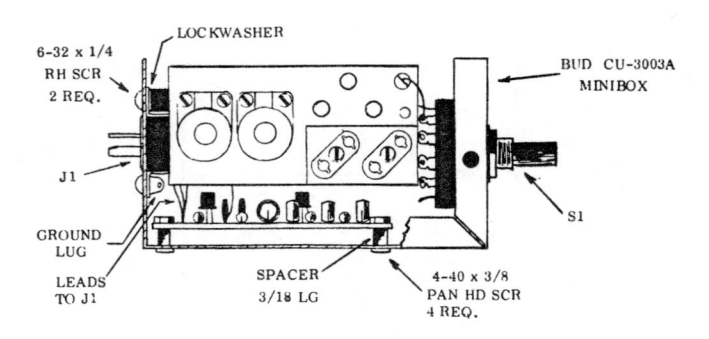

Figure 3

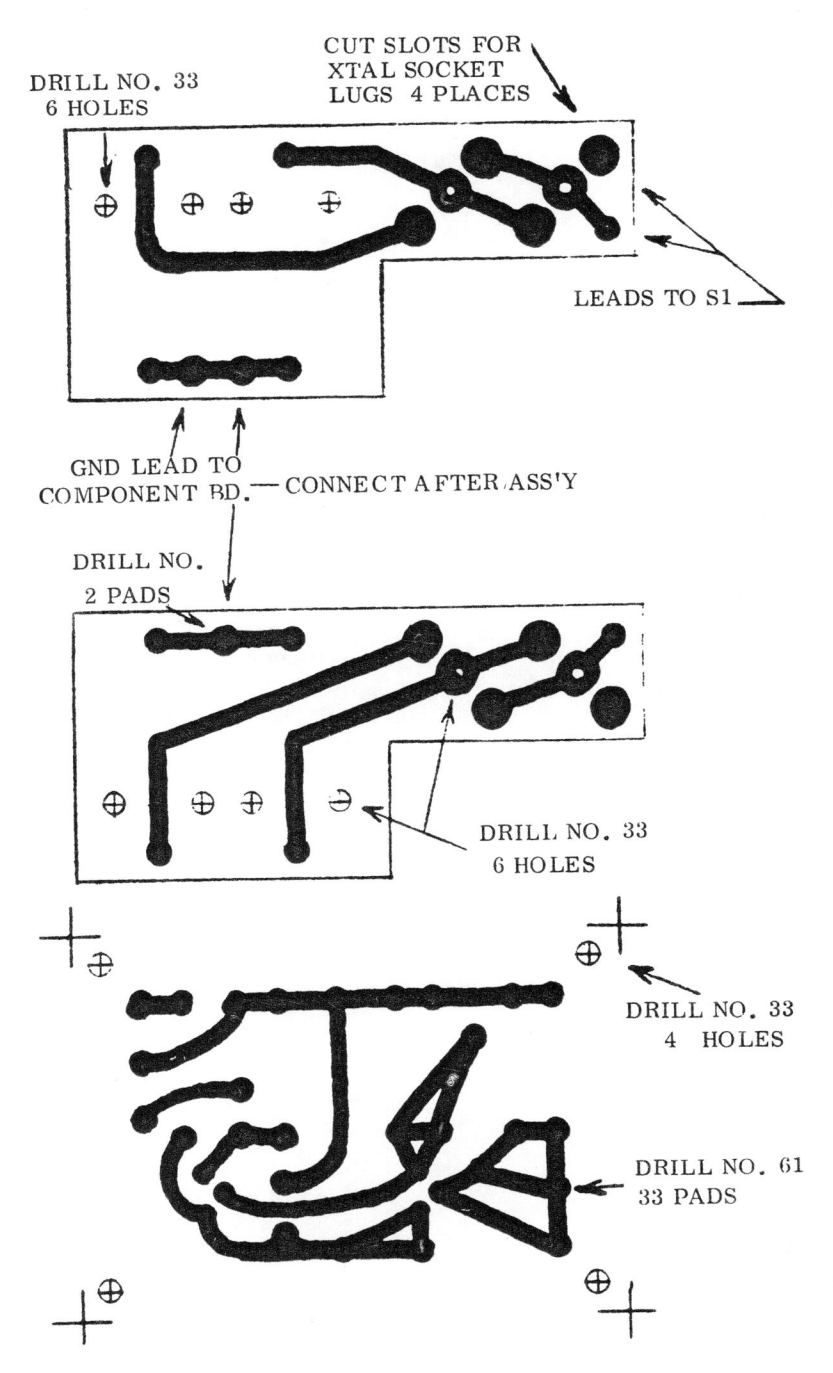

DRILL NO. 33
6 HOLES

CUT SLOTS FOR
XTAL SOCKET
LUGS 4 PLACES

LEADS TO S1

GND LEAD TO
COMPONENT BD. — CONNECT AFTER ASS'Y

DRILL NO.
2 PADS

DRILL NO. 33
6 HOLES

DRILL NO. 33
4 HOLES

DRILL NO. 61
33 PADS

53

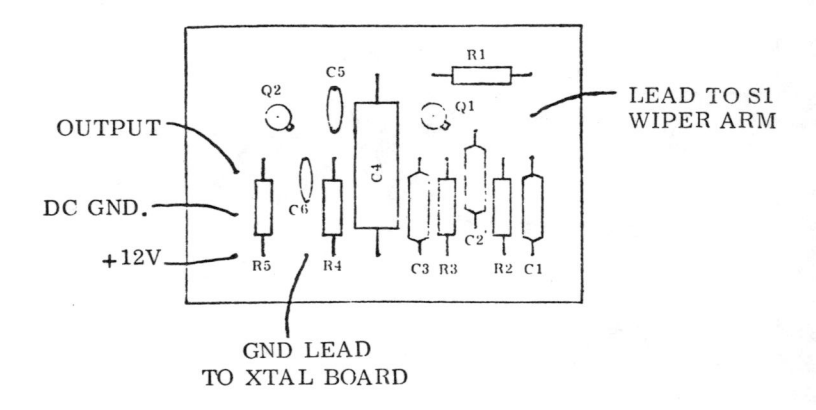

used for mechanical stability. Use care when inserting or removing the crystals so as not to crack the circuit boards. Glass-epoxy board material is worth using, if available, because of its greater strength.

When adjusting the trimmers for exact frequency, remember that placing the cover on the box will lower the frequency slightly (10 to 20 Hz with 3 MHz crystals) due to the added capacitance. To avoid this problem, holes can be drilled in the cover for access to the trimmers. Because of its small size, the oscillator can easily be taken to a counter or other measuring equipment for accurate frequency adjustment.

Measurements of thermal stability have not been taken yet, but experience has shown that in the usual mobile environment the frequency does not change enough to be evident at the receiving end.

If you have any trouble making the remote oscillator work, please get in touch with the author.

Dial-On
Frequency Standard

The fellow who first said "Necessity is the mother of Invention" might well have been an FM'er. What remote repeater operator hasn't found it necessary to traipse up to the site for no better reason than to zero in a few user mobiles? In my case, the combination of distance (25 miles to the site) and increasing repeater population (12 users) prompted me to seek a method for accomplishing this function without physically viewing a discriminator meter.

The challenge was posed to our more technically inclined members, and within a short time Bob Mueller (K6ASK) came up with the perfect solution. He suggested installation of an oscillator at the site that could be dialed on from the control points. The oscillator, crystal controlled at the discriminator frequency, is coupled, on command, into the low i-f of the repeater receiver so the receiver "thinks" it sees a weak on-channel signal. With the oscillator in the circuit, mobile operators with duplexed radios need only zero their transmitters while monitoring the heterodyne of the output for a zero indication.

Since our remote receiver is a GE Progress Line, the crystal oscillator frequency was 290 kHz. In the event you don't feel inclined to construction, Sentry Manufacturing Company sells the complete oscillator assemblies as well as the crystals. The units (less crystal) cost about $5.00 each.* The Sentry catalog shows various simple oscillator circuits if you'd rather build your own as Bob did. Figure 1 shows an oscillator circuit that is compatible with a 28-volt control system.

When you install the oscillator, disconnect the repeater receive antenna during setup. Then, with the oscillator in the circuit, tune the dial-on oscillator trimmer for zero on the discriminator, and reconnect the antenna.

If you don't take certain precautions to maintain control of your repeater, you'll likely experience the same sad fate we did: that of being jammed out by our own brainchild. When the

* Sentry Model F60-112.

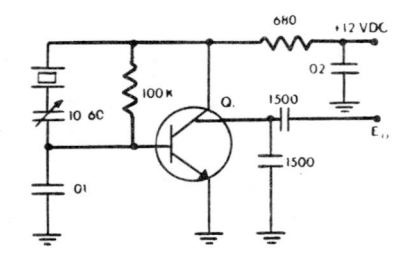

Figure 1

oscillator is dialed on, the oscillator feeds the receiver a legitimate signal. And if your signal is weaker than the oscillator's, you've lost control. This problem can be precluded by installation of a timer to turn off the oscillator automatically after a predetermined period (30 seconds or so).

Figure 2 shows how this can be accomplished using a single pulse for control without a latching relay. When the function is dialed on, a −28 volt pulse pulls in the relay, which stays closed because its own contacts supply coil voltage. The timer (an Agastat pneumatic type for best results) begins the timing sequence the instant the relay pulls in.

The normally closed timer contacts supply the control voltage for the oscillator as well as the constant relay voltage. At the end of the timed period, the timer contacts open, releasing the relay and removing control voltage from the oscillator.

The diodes may or may not be required in your own control scheme, but it is good engineering practice to incorporate them. Diode D1 prevents the trigger pulse from working too hard; it is arranged so that presence of a negative pulse keys only the relay. Remote owners who don't use diodes in circuits such as this sometimes find themselves with

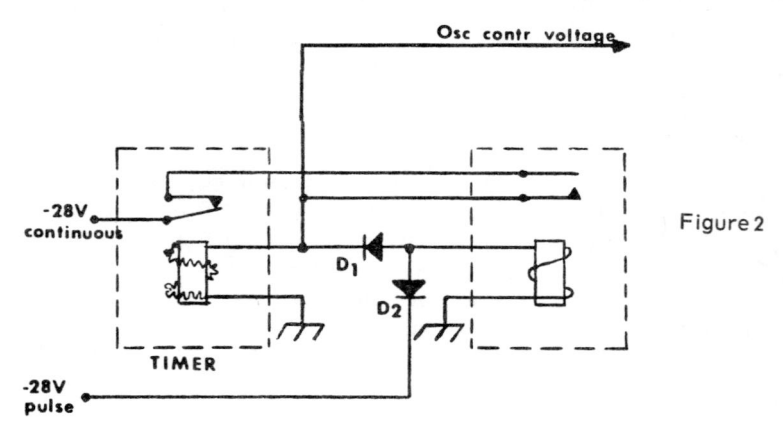

Figure 2

unreliable control systems. When a pulse is required to perform a number of functions, the excessive current can cause the pulse voltage to drop to such a level that many functions may go untriggered. Diode D2 offers nothing more than isolation; it prevents the constant relay coil voltage from appearing at the control wiper arm or the pulse source.

Controlled Charging of Ni-Cad Batteries

A desirable feature of a ni-cad charger is the capability of supplying a diminishing charge to the battery to minimize the gassing characteristic of ni-cad cells during the final charging hours. Motorola's approach is both effective and easy to duplicate, as shown in the circuit.

The capacitor in the circuit is selected according to the charge rate required, which should be on the order of 10 percent of the ampere-hour capacity. In the case shown, the ideal battery to be charged would be the 4 ampere-hour types used in Motorola's P33 and P31 units. The 8 uF capacitor drops the initial charge current to approximately 375 mA.

The No. 43 lamp serves several functions. It is fuse for the entire circuit and it acts as a charging indicator. But most important, it is a current-limiter. The No. 43 lamp draws approximately one-half amp. By replacing it with higher- or lower-rated lamps, the charge rate can be controlled to some degree.

If the charge rate is to be changed to any great extent, the capacitor value should be changed. The proper capacitor for charging the 250 mA-hr cells in the Handie-Talkie series of units would be 0.5 uF.

The lamp exhibits the greatest degree of brilliance when the charge cycle is initiated. Toward the end of the charge, the lamp dims somewhat, indicating charging is complete.

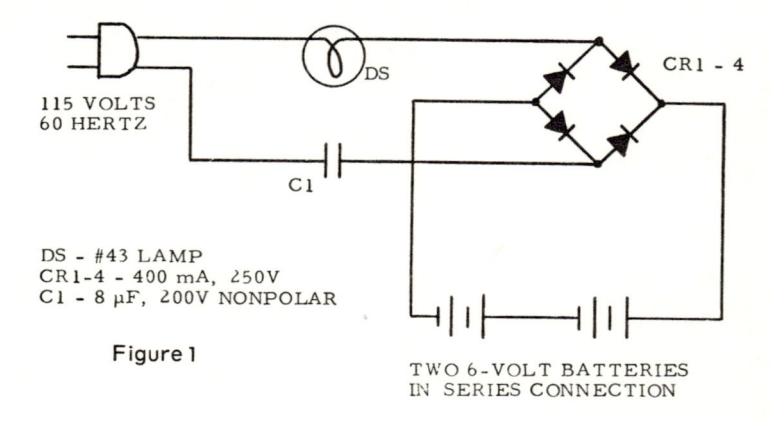

115 VOLTS
60 HERTZ

DS - #43 LAMP
CR1-4 - 400 mA, 250V
C1 - 8 µF, 200V NONPOLAR

Figure 1

TWO 6-VOLT BATTERIES
IN SERIES CONNECTION

Integrated Circuit Repeater Identifier

by Tom Woore

There is no economically adequate way of identifying a repeater automatically. The FCC rules require that an amateur repeater be identified every three minutes that the system is in use. Though some systems disregard the rule, others are identified via voice by each person using the repeater. The more sophisticated systems identify automatically by employing mechanical code wheels or relays, both of which have a high mortality rate. The question comes to mind, why not a better mousetrap? Or in this case, why not a solid state or more state-of-the-art integrated circuit identification unit!!

With no moving parts and a parts cost of about $18, this digital identification unit can be built to outlast anything you put on top of a mountain.

The digital identification unit (DIU) is unique in that it uses a simplified computer address principle for selecting the information it is programed to send. There are three basic units in the DIU: counter, matrix (memory), and signal logic. The counter establishes which sequence is next. The matrix determines what instruction is next by the sequence. The signal logic converts the instruction information into the actual signal to be sent. The whole system is based on a closed loop and therefore no standard clock is employed in the logic.

To understand how the DIU works we must first become familiar with some of the simple logic units that the system is based on.

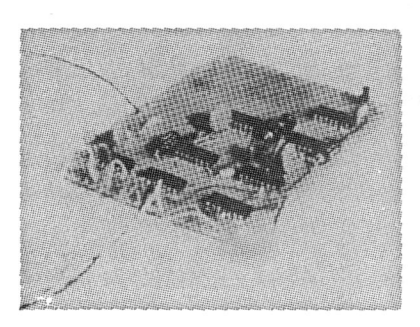

High: Output of logic unit (at least +1.5 volts).
Low: Output of logic unit (less than 0.5 volt).
Inverter: Device used to produce opposite logic state of what is fed into it. Example: +2 volts input into an inverter would produce a zero output while a zero input would produce a +2 volt output. Symbol:

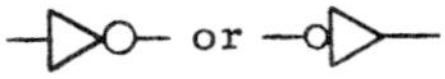

$$\text{—}\!\triangleright\!\circ\text{—} \quad \text{or} \quad \text{—}\!\circ\!\triangleright\text{—}$$

OR gate: Device used to give an output for any of the signals fed into the input. Example: 3 inputs, one at +2 volts, the other two at zero, would give a +2 volt output. Symbol:

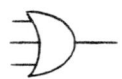

AND gate: Device used to give an output for all input lines being high. Example: +2 volts on all 3 inputs of gate produces +2 volts on the output. Symbol:

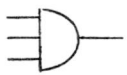

NOR gate: An inverted OR gate; device to give a zero output when any of its inputs are high. Example: 3 inputs, one at +2 volts, the other two at zero, would give a zero output. Symbol:

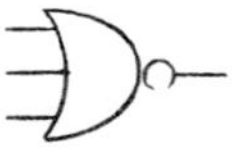

NAND gate: An inverted AND gate device used to give a zero output when all of its inputs are high. Example: 3 inputs all at +2 volts produces an output of zero volts. Symbol:

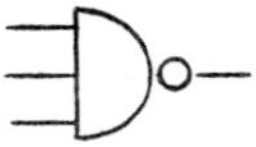

For this article, NOR gate logic was used for NAND functions; therefore, the definition for our purpose of a NAND gate is a device used to give a high output when all of its inputs are zero. Example: 3 inputs at zero produces +2 volts output. Symbol:

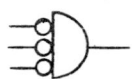

Note that the zero placed before or after the inverter, NOR, and NAND logic units defines the expected state of the input or output.

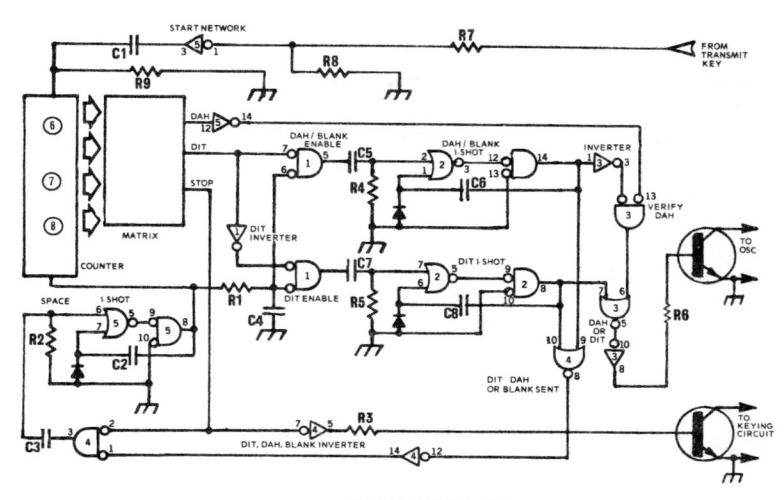

PARTS LIST

R1, R7	3.3K
R2, R3, R4, R5, R6, and R9	10K
R8	6.6K
C1, C3, C4, C5, and C7	.05 uF
C2, C8	10 uF
C6	30 uF
Diodes (including matrix diodes), 20 to 100	
1, 2, 3, 4, 5 (within symbols above)	HEP 570 (Motorola)
6, 7, 8 (within symbols above)	HEP 572 (Motorola)
Transistors	2N3415 (GE)

All resistors: one-quarter watt or greater capability.
All capacitors: 6 volts or greater, working capability.
Numbers outside symbols refer to pin contacts on IC's.
Grounds not shown.
Ground contacts are as follows:

IC NUMBER	GROUND PINS
1	12,
2	12, 10
3	2, 9
4	13, 6
5	10, 13, 2

Fig. 1. System schematic diagram, digital integrated circuit repeater identification unit.

Unit: Smallest bit of information sent by the digital identification unit, dit, dah, and blank.

The digital identification unit uses the MC700 series integrated circuits due to their inexpensiveness and availability. The Motorola HEP line of integrated circuits can also be used.

DIGITAL IDENTIFIER UNIT

A zero volt signal through the start network from the transmitter keying circuit resets all the flip-flops in the counter to the zero state. All Q lines become positive. The positive signals, approximately 2 volts, are fed into the diode matrix which decodes the counter number into an instruction for the oscillator keying logic. In the digital identification unit there are four basic instructions: (1) send a dit, (2) send a dah, (3) send neither dit nor dah (blank), and (4) stop.

If the diode matrix decodes the first sequence count (0) to be instruction no. 1 (send dit), the dit signal line from the matrix will be high. This will cause the dit inverter to have a low output and one-half of the dit enable NAND gate would be enabled. Since the space line is also at zero level at this time, a trigger pulse would be sent through capacitor C7 to the dit one-shot. (A one-shot is a monostable device used to generate a predetermined pulse width.) The dit time pulse determined by the one-shot is sent through the dit/dah NOR gate and inverter to key the tone oscillator circuit.

At the same time the dit is being sent by the one-shot to the oscillator, the space one-shot logic is being reset ("dit, dah, or blank," NOR gate, "dit, dah, or blank," inverter, and "space enable" gate).

Upon completion of the dit signal, the "dit, dah, or blank" NOR gate output becomes high making "dit, dah, blank" inverter output low. Since the stop instruction has not been called for by the matrix, the space enable NAND gate produces a high output. The high output from the space enable NAND gate sends a pulse through capacitor C3 to trigger the space one-shot. (The space time period is used to separate the units of a letter. Example: D = dah space dit space dit.)

The space period is the same as the period for a dit. The space signal, besides allowing for the time to distinguish the units of a letter, advances the counter to the next unit and resets the dit and dah one-shots by discharging capacitors C5 and C7.

If the diode matrix decodes the next sequence to be instruction no. 2 (send dah), the dah signal line from the matrix will be high and the dit signal line will be low. When the space line becomes low, the dah/blank enable NAND gate will send a pulse through capacitor C5 triggering the dah/blank one-shot. The dah time pulse would then go through the dah/blank inverter and be challenged by "verify dah" NAND gate to see if the pulse was for a dah. An affirmative check would come from the dah inverter with a zero level output. The dah signal would then be sent through the dit/dah NOR gate, inverted, and sent to the oscillator keying circuit. The space one-shot is then triggered to advance the counter to the next unit.

If the diode matrix decodes the next sequence to be instruction no. 3 (send a blank), neither the dah nor dit line will be positive. The same will occur again as if a dah were being sent, except that when the signal reaches the "verify dah" gate, it will be stopped from keying the oscillator. This generates the blank period which is put between letters. (DE = dah space dit space dit blank dit blank.) Again the loop through the space one-shot is triggered and the counter is advanced to the next unit of information.

The counter is advanced each time a unit of information is sent until it is advanced to the "stop" instruction. This instruction causes a blank to be automatically sent and stops the space enable NAND gate from triggering the space one-shot. The digital identification unit remains in the stop state until a reset pulse is sent to the counter from the transmitter keying circuit again.

The R1, C4 network is used to slow down the fall of the space line so that the counter is allowed to advance before the sending logic is ready to send the next unit of information. Transistor T2 is used to key the transmitter keying circuit while the digital identification unit is sending its identification code.

DIODE MATRIX

Up until now very little has been said about the diode matrix other than the fact that it determines what instruction to give the keying logic. The actual construction of the matrix can be made considerably cheaper by simplifying the diode logic. Up to 70 percent of the diodes necessary for the diode matrix can be eliminated by using mathematics. A much more sophisticated, economical, and space-saving layout can be achieved using Boolean algebra. The matrix in this DIU employs the simplifying techniques of Boolean algebra.

Thanks to Mr. Karnaugh, it is not necessary to give a complete discussion on Boolean algebra. The Karnaugh map, Fig. 2, is a device for mechanically determining the mathematical equivalent of the diode matrix. For the purposes of this discussion the MCW message for the DIU will be

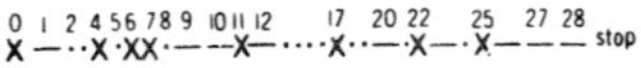

```
0 1 2 4 5 6 7 8 9 10 11 12   17  20 22   25  27 28
X — · · X ·XX· — — X— · · · ·X· — X — X— — — stop
```
Fig. 2. Identification unit breakdown.

"DE W6FNO." Of course, any other message can be developed by this method and consequently this discussion may be used for developing any matrix logic for controlling a system.

The first step in determining the diode matrix for the message is to break up the message into the units to be sent. (. = dit, — = dah, x = blank). This is shown in the breakdown diagram of Fig. 2.

It is seen that 30 units of message will be sent (0 is actually used for a blank). To convert units 0 to 29 into information as to how many diodes will be used, the Karnaugh map is needed.

The numbers in the box (Fig. 3) correspond to the decimal number equivalent to units on the output of the counter. The numbers across the top and along the side of the chart correspond to the binary output of the flip-flops —1 for true or 0 for false. The letters written diagonally refer to the five flip-flops. Example: Box 17 has flip-flop A true, B false, C false, D false, and E true. Written in Boolean form, 17 would be represented by $A\bar{B}\bar{C}\bar{D}E$, where the bar over the letter means that the flip-flop is false and, conversely, a letter without a bar (CBA) represents a flip-flop that is true.

To simplify the matrix, a Karnaugh map is constructed for both the dits and dahs to be sent. From Fig. 2, 2, 3, 5, 8, 13, 14, 15, 16, 18, 19, 21, and 24 represent the dits to be sent in the message. In the dit Karnaugh map (Fig. 3) a "1" is placed in each box corresponding to the number. An X (not the x which represents a blank) is placed in all boxes after the stop code number. These are "don't care" conditions because the counter will not count to these codes.

From the dit Karnaugh map it can be seen that the third unit of information is a dit and that the flip-flop A is true, B is true, C is false, D is false, and E is false (or $AB\bar{C}\bar{D}\bar{E}$). To put this in matrix form, the Boolean algebra tells us that this dit would be represented by a diode connected to Q_A lead (the true lead of flip-flop A), another to Q_B, another to Q_C (the

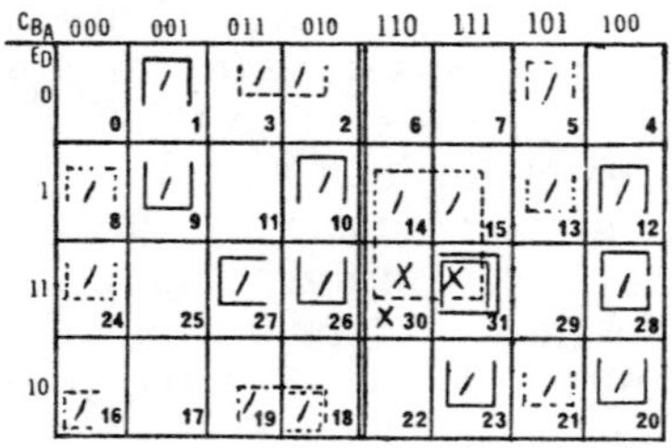

Fig. 3. Karnaugh map (dit = ··· dah = —)

64

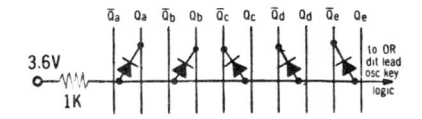

Fig. 4. Unit 3 information dit.

false lead of flip-flop C), another to Q_D, another to Q_E. It would take five diodes normally for sending this information (Fig. 4).

It would normally take five diodes for each unit of information in the message or 29 x 5 = 145 diodes. This does not count the diodes needed to OR the dahs together and dits together, which requires an additional 21 diodes. A total of 166 diodes would be used. This is where the Karnaugh map saves diodes. On the map (Fig. 3) any adjacent box or any box that changes just one variable from another box eliminates that variable. Boxes 8 and 24 simplify to $\bar{A}\bar{B}\bar{C}D$, eliminating the E flip-flop altogether. Boxes 3, 2, 19, 18 also simplify since they change one variable at a time (or $B\bar{C}\bar{D}$). Note that not only is 3 (Fig. 3) represented by $B\bar{C}\bar{D}$, but 2, 19, and 18, resulting in a savings of 20 — 3 = 17 diodes. 14, 15 combine with "don't cares" 30, 31 to equal BCD. The final expression for the dits is $\bar{D}\bar{C}\bar{C} + BCD + \bar{A}\bar{B}\bar{C}\bar{D} + \bar{A}\bar{B}\bar{C}\bar{E} + \bar{A}\bar{B}\bar{C}E + A\bar{B}\bar{C}\bar{D}$. A total of 22 + 6 (number of OR groups) = 28 diodes are used to express all the dits in the message. A total of 61 diodes makes up the complete matrix including dits, dahs, and stop codes. Quite a few less than 166!

The final matrix appears in Fig. 5 for the message DE W6FNO. Note that any matrix of this magnitude can be determined by the above method. Figure 5b shows wiring for the counter. Note that Fig. 5b mates to the leads of Fig. 5a.

CONSTRUCTION

The layout used for the identifier was adopted so that the unit could be mounted to the side of the transmitter cabinet of the repeater on standoffs. However, any physical configuration can be used to develop the layout.

If sockets are not used for the ICs, care should be taken so that they are not overheated. A small "pencil" iron will do nicely for soldering the ICs rapidly to the PC board.

All parts are readily available from most electronic parts houses. The ICs and the 3.9V zener used in the project were obtained from Hamilton Electrosales. The power supply (Fig. 6) was designed to be shortproof and limits current to a maximum of 1 amp at 3.6 volts.

INSTALLATION

The signals normally received and sent to and from the DIU should meet the following criteria:

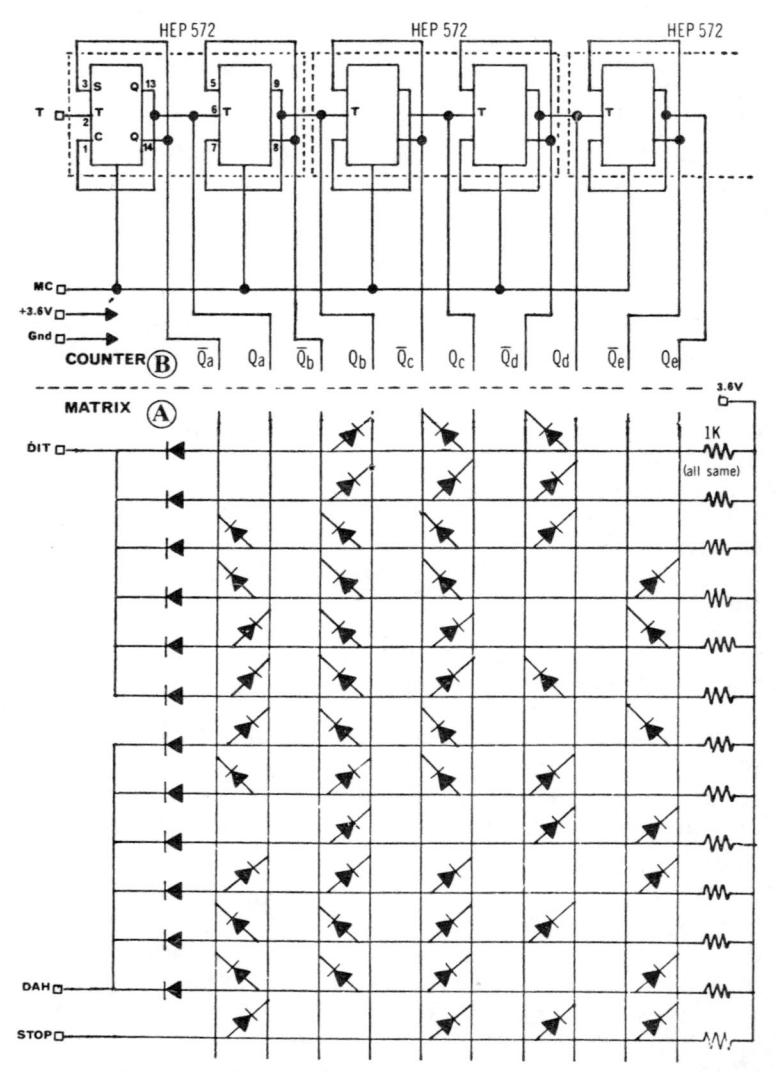

Fig. 5. "DE W6FNO" board & universal counter.

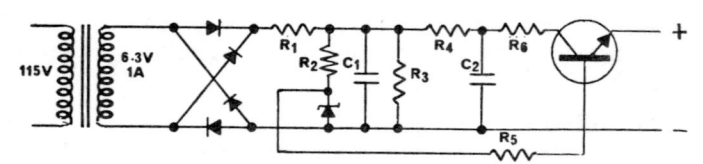

Fig. 6. Regulated IC power supply.

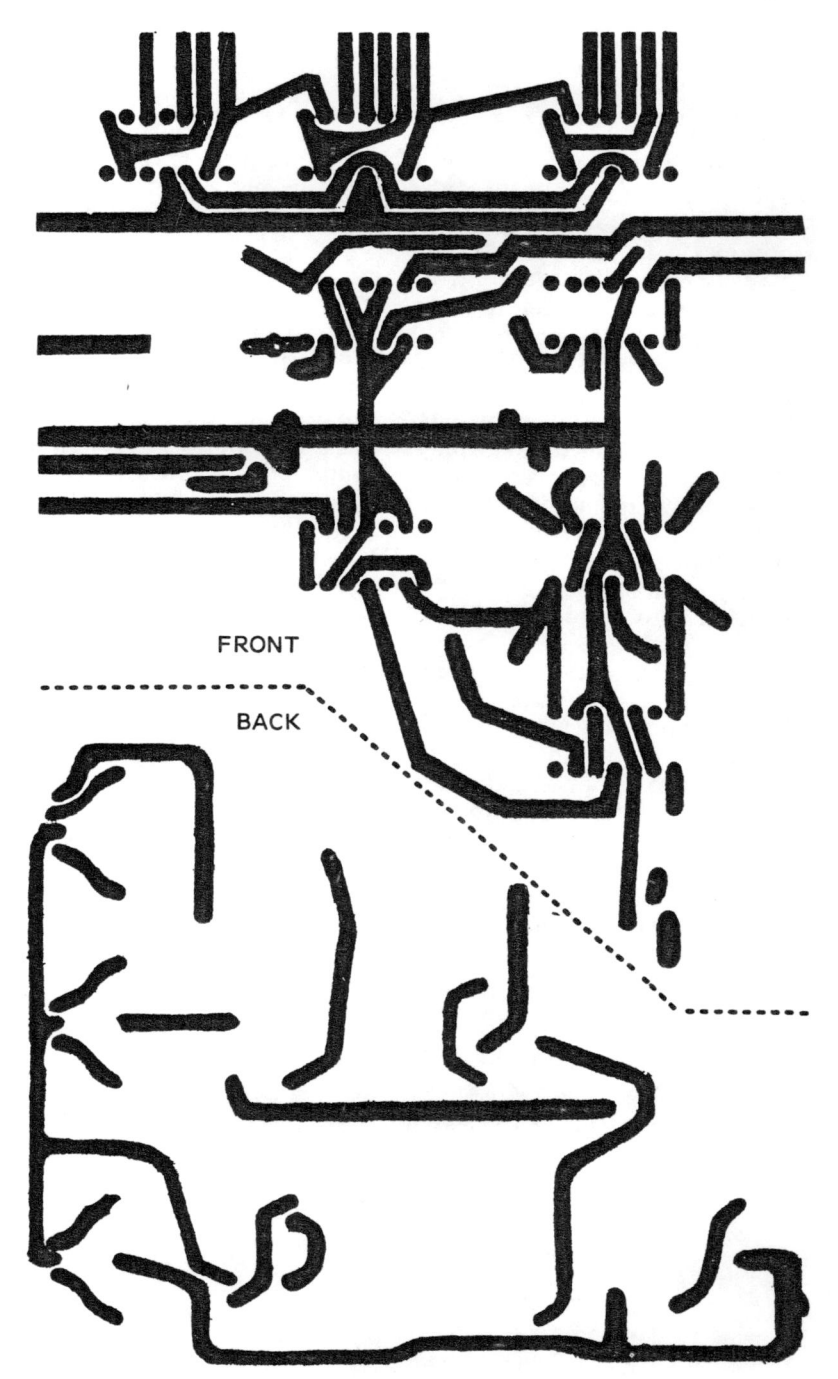

Fig. 7. Circuit board layout (2 sides).

1. From power supply: 3.6 volts dc, well filtered and regulated.

2. From transmitter keying circuit: 0 volt transmitter keyed; approximately 6 volts transmitter not keyed.

3. To oscillator keying circuit: tone off, 10 megohms (nominal); tone, 10 ohms.

4. To transmitter keying circuit (identifier off): 10 megohms (nominal); (identifier on: 10 ohms).

Note that all lines should be filtered dc. In some relay circuits the output of a bridge rectifier is used to directly key the transmitter. In order that the pulsating dc does not key the digital identification unit, a 60 uF capacitor or greater should be put across the relay supply. While installing the W6FNO DIU, it was discovered that the grounded 6.3 volt filament supply in the transmitter could not be used to power the DIU. The reason this arrangement could not be used was because the output of the rectifier in the DIU was grounded. If any of the signal lines do not meet the criteria set above, simple diode, resistor, and capacitor circuits can be used to condition the signal.

When the final installation of the DIU was completed for W6FNO, the transmitter keying circuit from the DIU was disconnected; this was due to the fact that the DIU only takes 2 seconds (42 wpm) to identify the station. If the carrier keying the repeater dropped out while the DIU was identifying, the completion of the MCW would be lost in the squelch tail of the receiver.

Obviously, the PC board for the W6FNO matrix could not be used for other systems. The DIU, however, is applicable to all identification systems. Actual-size patterns for the board (two sides) are provided in Fig. 7 (a and b).

SUPPLY PARTS

RESISTORS:

R1,4 1 ohm, half watt	2
R6 8 ohm, 2 watt	1
R3 1K ohm, half watt	1
R2 220 ohm, half watt	1
R5 100 ohm, half watt	1

CAPACITOR:

200 uF, 15 volts	2

SEMICONDUCTORS:

HEP 245 (or 2N4921) transistor	1
HEP 175 bridge rectifier	1
1N5228 zener, 3.9V, 0.5W	1

Telephone Command of Repeater Operations

A not-too-often considered means of remote control is the telephone itself (assuming a telephone is available at the remote site). As a principal control element, the telephone has certain disadvantages, but as a backup system the telephone has no equal. There is no feeling quite as comfortable to a remote or repeater owner as the knowledge that he can shut down his system regardless of what happens to the hilltop transmitter or receiver and regardless of where he is. For he knows that to accomplish shutdown, he need only go to the nearest telephone and dial the remote number. When the remotely situated telephone rings, the shutdown function will occur.

The circuit of Fig. 1 shows how the telephone ringing voltage can be used to trigger a selected function without causing interference to the phone line. The ac ringing voltage is isolated from the phone lines through C1 and C2 and rectified to produce a dc signal which triggers the current-operated relay. Omission of C1 and C2 would cause excessive loading of the phone line and would result in hum, level problems, and dc entry. The filter capacitor must be low enough in value to allow full charging during a one-second ring so that enough power is available to pull in the relay.

It is easy to see the difficulties that could arise if the telephone number were commonly known, since any ring

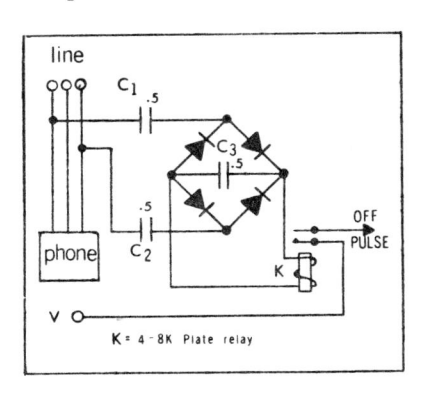

Fig. 1. Safety "off" system.

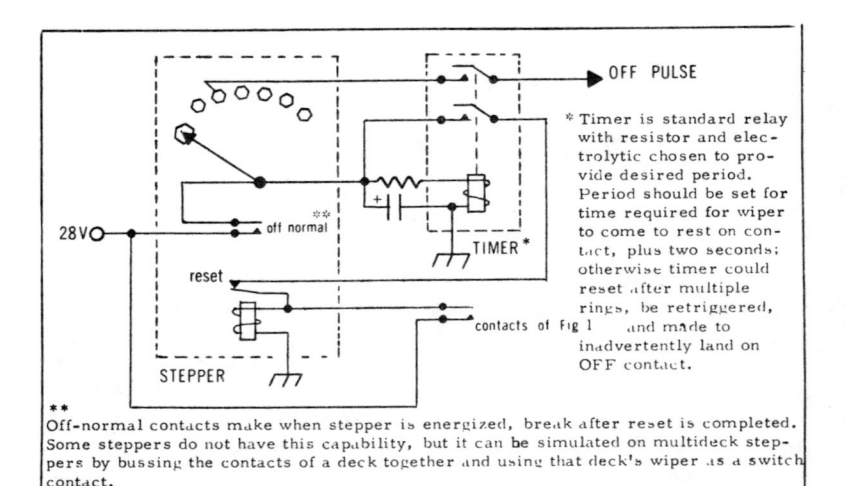

```
                                                    ► OFF PULSE
                                    * Timer is standard relay
                                      with resistor and elec-
                                      trolytic chosen to pro-
                                      vide desired period.
                                      Period should be set for
              ** off normal           time required for wiper
 28V○────────                         to come to rest on con-
                                TIMER* tact, plus two seconds;
         reset                         otherwise timer could
                                       reset after multiple
                                       rings, be retriggered,
                                contacts of Fig 1  and made to
                                       inadvertently land on
                                       OFF contact.
         STEPPER

**
Off-normal contacts make when stepper is energized, break after reset is completed.
Some steppers do not have this capability, but it can be simulated on multideck step-
pers by bussing the contacts of a deck together and using that deck's wiper as a switch
contact.
```

Fig. 2. Coded "off" pulsing scheme.

would cause immediate interruption of repeater service. This problem can be avoided by adding the extra circuitry required so that the system will shut down only when the phone rings a specific number of times. This circuit is shown in Fig. 2

In the case shown, the desired function occurs when the phone has rung exactly twice. The first ring prevents the stepper from reaching the right point. If a third ring occurs, the function is canceled. The function occurs only when the phone has rung two times and the stepper rests on position 2. Ringing of the phone energizes the timer and moves the stepper. The stepper will automatically reset after it has been energized. The period of the timer should be selected to allow sufficient time for the stepper wiper to come to rest on the "off" contact with a few seconds to spare.

Quickie Tone Generator...
for Whistle-On Use

In many ways, the W6FNO repeater at Radio Ranch could serve as a model installation. The repeater stands ready for use 24 hours a day and is never shut down where it cannot be accessed by a station on the input frequency (146.82 MHz). On the other hand, the repeater will shut itself off if a three-minute period elapses with no signals on the input. Sounds a little contradictory, but it isn't—not really. The W6FNO repeater was an experiment to test the concept of subcontrol, i.e., limited control of the repeater from the actual frequency of operation.

The FCC sanctions use of semicontrol techniques such as continuous-tone squelch and single-tone in applications requiring limited access to a remote or repeater. The W6FNO repeater crew took the idea one step further, and the result is a repeater that is fully compatible with on-channel nonrepeater operation, and one which does not pointlessly add to the congestion of a crowded band.

The repeater is equipped with two timers. The first timer is a transmission-limiting device: when the input carrier exceeds three minutes duration, B-plus is removed from the transmitter final; and it can only be reapplied after the input carrier drops out momentarily. The second timer removes the transmitter B-plus also. But in this case, it is activated from absence of an input signal for three minutes. Since the shutdown is only B-plus removal, the repeater stands ready to be activated immediately upon application of the proper signal, which in this case is nothing more than a shrill whistle.

The active FM channel in the W6FNO area is the input frequency, 146.82 MHz. The repeater output frequency is not used at all except to monitor the repeater output. When two stations are conversing on the FM channel, the repeater is not even part of the operation unless one of the operators wants it to be (as for instance, when the copy gets rough).

When a user wants to monitor the active 146.82 channel, but he is too far away from the area of activity to hear the stations, he merely puts a carrier on 146.82 and whistles into the microphone. Instantly the repeater comes on, regardless of the time of day or night, and the user finds himself right in the middle of the action. The only difference is that he hears his .82 on 146.70.

The decoder at the repeater site that provides the turn-on function is nothing more than a simple frequency-to-dc converter such as the semiconductor decoder described elsewhere in this book. This device is set to respond to as broad a range of frequencies as possible without being energized from voice tripping.

Even though the W6FNO decoder was set to respond to a wide frequency range, a few users found it difficult to key the repeater on. Perhaps their audio was not quite what it should be to reproduce the required tone (1750 Hz), or perhaps they were simply not proficient at whistling. At any rate, one of the users (W6ZCL) hit upon the idea of installing a simple automatic whistler in each of his transmitters.

The circuit he designed was too uncomplicated to be considered an encoder; it consists of nothing more than a single-transistor oscillator using a twin-T feedback network. As can be seen from the circuit of Fig. 1, the design is the absolute epitome of simplicity—and it works every time.

The whistler was not connected so that it would go on with each transmission; not only would this have defeated the purpose of the automatic-off function, but it would have given him the unpleasant characteristic of a squeal at the outset of each transmission. Instead, he connected the device into his unit so that it is energized by pressing a momentary-contact switch on the control head. Figure 2 shows how the oscillator is used with a Motorola unit.

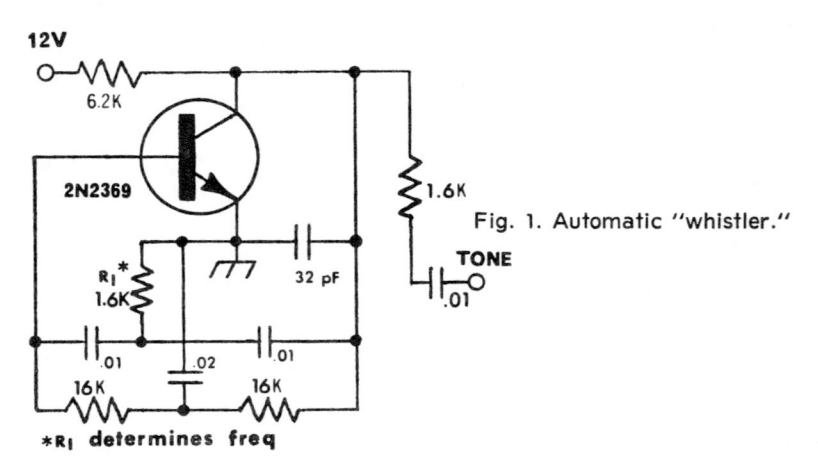

Fig. 1. Automatic "whistler."

*R_1 determines freq

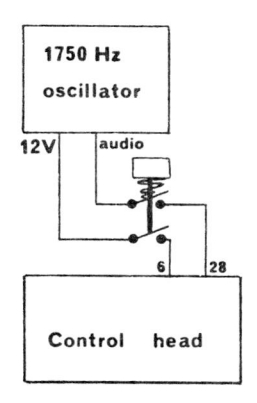

Fig. 2. Interconnection.

The automatic shutdown feature of the W6FNO repeater has done a great deal to enhance the relationship between the repeater users and the nonrepeater users in its locale. That segment of amateurdom that is against repeaters because they are used without necessity on many occasions can be satisfied that the repeater "responsibles" are doing their part to minimize the likelihood of unnecessary operation and use, but to provide an unmatched communications capability when it is needed.

The Peaker-Tweaker
by P. J. Ferrell

Once upon a time, in those days before the Seattle repeater, rigs had to be kept in tip-top shape for satisfactory "direct" communications. Nuvistor and 417A preamps were the rule rather than the exception. Those whose receivers took more than 0.3 uV for 20 dB quieting were quietly pitied or openly ridiculed, or both. To keep receivers at maximum sensitivity, tuneups were performed at the drop of a hint. Because of the scarcity of "official" signal generators, a crystal-controlled one-transistor signal generator was developed. This gadget became known as a "peaker-tweaker." Seattle has a two-meter repeater now, so no one bothers much with his receiver any more, but one still remembers those good old days of superceivers and how out-of-towners with "dead receivers" took plenty of gas from the local troops.

The peaker-tweaker is shown in Fig. 1. As a signal generator, the device has everything going for it. It's easy to build, simple in design, and straightforward in theory; yet, it provides a clean, stable signal that can be used in tuneup of receivers on any of the VHF or UHF bands.

Here's how the peaker-tweaker works:

Inductor L1 doubles the VXO action of C1, allowing a single crystal to cover several adjacent channels. L1 may be

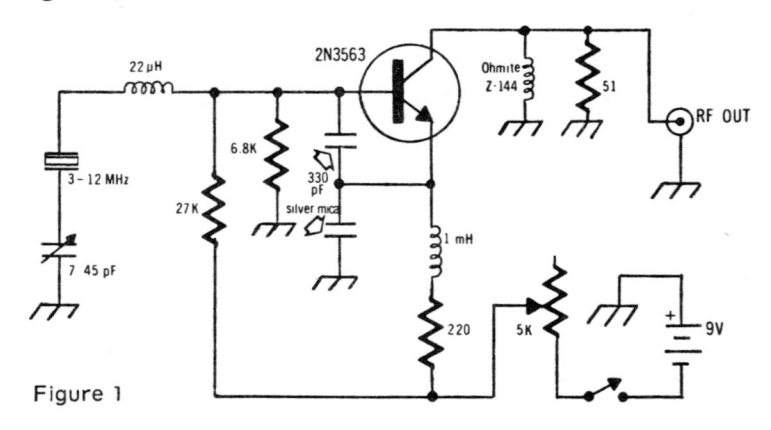

Figure 1

Harmonic Number	Crystal Frequencies in kHz 146.34	146.76	146.94	Harmonic Number	Crystal Frequencies in kHz 146.34	146.76	146.94
40	3658	3669	3673	27	5420	5436	5442
39	3752	3763	3768	26	5628	5645	5652
38	3851	3862	3867	25	5854	5870	5878
37	3955	3966	3971	24	6098	6115	6122
36	4065	4077	4082	23	6363	6381	6389
35	4181	4193	4198	22	6652	6671	6679
34	4304	4316	4322	21	6969*	6989	6997
33	4435	4447	4453	20	7317	7338	7347
32	4573	4586	4592	19	7702	7724	7734
31	4721	4731	4740	18	8130	8153	8163
30	4878	4892	4898	17	8608	8633	8644
29	5046	5061	5067	16	9146	9172	9184
28	5226	5241	5248	15	9756	9784	9796

*not often available

omitted entirely if the additional coverage is not desired. The rf output is a series of very fast pulses at the crystal frequency. Such a waveform is extremely rich in harmonics.

Regular transmitting crystals may be used, but any crystal in the range of 3-12 MHz works as well. The transistor specified is a VHF type which can put out up to —20 dBm into a 50-ohm load for harmonics up to 500 MHz. Oscillator pulling by the potentiometer (which controls output level) is minimal, about 1 kHz at high-band. Drift is just about nonexistent—a fact to be appreciated by those using most "regular" signal generators. To aid those with a selection of surplus crystals, a simple computer program was formulated to generate the following table of crystal frequencies (all of which have harmonics on several popular frequencies on two).

The Poor Man's Frequency Meter

Yes Virginia.

There IS a way to accurately measure frequency without the use of expensive test equipment. With a handful of parts and a few items found commonly around the radio shop, you can build the poor man's frequency meter, which is capable of tolerances that should amaze those of you who think two grand is what it takes to make the "trek to accuracy."

This is not a substitute for a good frequency meter for commercial use, but if you have a limited number of frequencies that you want to be "dead on," this may be the answer. It will provide a handy "extra" unit for your two-way shop, perhaps freeing the Gertsch for use elsewhere.

A nice feature of the poor man's frequency meter is the fact that it puts to use that old noise spectrum demodulator (communications receiver, that is) which has undoubtedly been sitting around in a dark corner of your basement under piles of old magazines and discarded dynamotors. It also uses that old wideband FM receiver you've been hesitating to throw away.

The idea is not new; it has been used for many years for the Motorola station monitor and various other common applications. Basically, it is composed of four major units:

- Receiver converter with calibration oscillator constructed on a high-band front-end deck from a Sensicon A receiver chassis.
- Monitor receiver (any 150 MHz wideband receiver); a low i-f of 455 kHz is best.
- WWV receiver (here is where the old communications receiver comes in).
- Accessory items (hang a modulation meter on it).

The block diagram of Fig. 1 shows how the individual items of equipment are interconnected to form the frequency meter. Note that although a narrowband receiver can be used, a more dependable "off frequency" indication is obtainable with a wideband i-f receiver. Periodic calibration (before use) to WWV is recommended for high-accuracy measurements; however, the unit will maintain its operating frequency to within 1 kHz (an error of 0.0006 percent) for an ambient temperature within the range of $-20^\circ C$ to $+60^\circ C$.

Figure 1

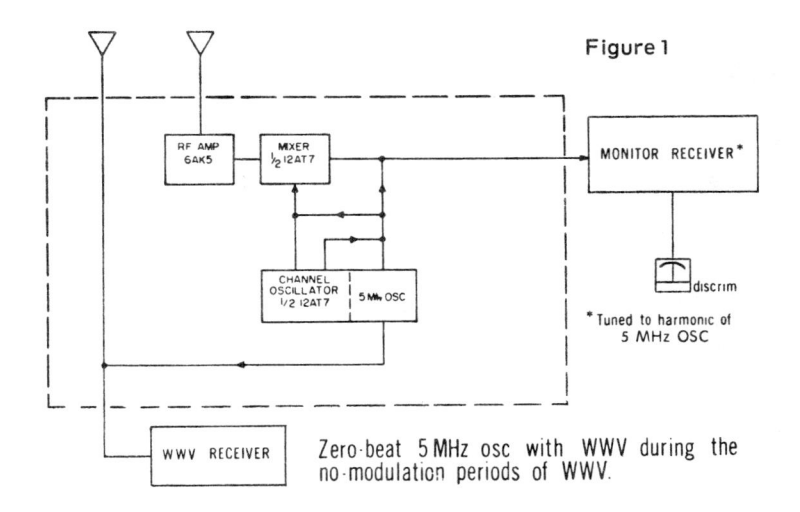

[Block diagram: RF AMP 6AK5 → MIXER ½ 12AT7 → MONITOR RECEIVER*; CHANNEL OSCILLATOR ½ 12AT7, 5 MHz OSC feeding into mixer; discrim meter shown below monitor receiver]

*Tuned to harmonic of 5 MHz OSC

WWV RECEIVER — Zero-beat 5 MHz osc with WWV during the no-modulation periods of WWV.

PRINCIPLE OF OPERATION

The frequency converter operates on the heterodyne principle. A station's frequency is monitored by heterodyning its carrier with the output frequency of the crystal oscillator and then feeding the resultant frequency of these two signals into the calibrated monitor receiver. If the beat frequency between the crystal oscillator signal and the monitored carrier is exactly equal to the frequency to which the receiver is aligned, the discriminator meter will indicate zero. If the beat frequency is lower or higher in frequency than the one to which the receiver is aligned, a direct indication of carrier frequency error in the monitored transmitter will be given on the meter.

The monitor receiver is aligned to a predetermined frequency. The specific frequency used will depend upon the spurious harmonics emitted by the channel crystals which will be required to monitor the specific carrier channels in consideration, plus the operating frequency to be measured.

The beat frequency fed to the control receiver may be either the sum or difference frequency of the channel crystal frequency and the monitored carrier frequency. Channel crystals for operation in the range from approximately 1.6 to 12.5 MHz may be used.

Calibration Oscillator

The calibrating oscillator consists basically of an rf amplifier stage, a mixer, and an oscillator. The calibrating crystal, shunted by a trimmer capacitor for any minor adjustment of oscillator frequency, is used for calibrating the monitor receiver.

Although the crystal is temperature controlled, a greater degree of accuracy is obtainable without the use of the heater. The trimmer capacitor provides exact calibration of the crystal frequency at any temperature by zero-beating the oscillator against the WWV signal.

The crystal heater should be used only when a quick check is necessary; such as, where it is desired to quickly bring the crystal to a temperature that would eventually be reached due to the heat dissipation of the equipment.

The control receiver may operate in the 145-160 MHz range; therefore, when using a 5 MHz calibration crystal, the 29th, 30th, 31st, or 32nd harmonic of the 5 MHz crystal frequency is used to calibrate the receiver to 145, 150, 155, or 160 MHz.

The selector switch operates in conjunction with the calibrating oscillator. This switch may be used to select any one of several crystals as the frequency controlling element of the oscillator. These crystals include the 5 MHz calibration crystal and the five channel crystals.

One half of a 12AT7 tube is used as the oscillator while the other half of the tube functions as the mixer. The carrier frequency to be monitored is picked up by the VHF antenna, amplified by the 6AK5 amplifier, and then mixed with the selected channel crystal frequency. The output of the mixer is fed to the calibrated monitor receiver where it is determined if the transmitted carrier is on frequency.

Assume that it is desired to monitor the output of the transmitter which is operating on 152.45 MHz and that the control receiver has been calibrated at 150 MHz. The necessary channel crystal frequency will be the difference between 152.45 MHz and 150 MHz, or 2.45 MHz. If the transmitter is on frequency, the 152.45 MHz signal will mix with the 2.45 MHz channel crystal frequency to produce an input signal of 150 MHz at the control receiver. In this case, no indication will be given by the discriminator meter. If the monitored transmitter carrier is above or below its designated frequency, the input signal to the monitor receiver will be above or below 150 MHz, causing the discriminator to produce an output voltage. This voltage is fed to the meter which is calibrated in kilohertz to give a direct reading of carrier frequency error.

When monitoring transmitters which operate in the 440-470 MHz band, the monitor must be placed so that the monitor antenna is within a few feet of the transmitter. For this application, the frequency of the stage preceding the final tripler is monitored. This is done by selecting a difference frequency crystal for the monitor which, when beat against the frequency of the transmitter stage preceding the tripler, produces the frequency at which the control receiver is tuned.

Assume that it is desired to monitor the output of a transmitter which is operating on 453.750 and that the monitor receiver is calibrated at 160 MHz. The channel crystal frequency is determined as follows:

$$453.750 \div 3 \text{ (tripler)} = 151.250 \text{ MHz (frequency actually monitored)}$$

$$160.00 - 151.250 = 8.75 \text{ MHz (channel frequency crystal)}$$

If the transmitter is on frequency, the frequency of the stage preceding the final tripler will mix with the 8.75 MHz channel crystal frequency to produce an input signal of 160 MHz at the monitor receiver. In this case, no indication will be given by the discriminator meter. If the monitored transmitter carrier is above or below its designated frequency, the input signal to the monitor receiver will be above or below 160 MHz, causing the discriminator to produce an output voltage. This voltage is fed to the discriminator meter which can be calibrated to give a direct reading of carrier frequency error.

Any error in carrier frequency indicated on the discriminator meter is an error in the frequency of the stage preceding the tripler; therefore, the error in the transmitter signal from the final amplifier will be three times as great. When using this method of monitoring, check the output (440-470) of the transmitter with a reliable wavemeter to ascertain that proper frequency multiplication is made.

Channel Crystal Accuracy

Since the fundamental frequency of the channel crystals is used, any error in crystal frequency is not multiplied. Therefore, the error in monitoring a frequency by this method is very small. Crystals are held to within 0.002 percent of the specified frequency over the ambient temperature range of $-30°C$ to $+60°C$. Therefore, with the previous example, the maximum frequency error of the 2.45 MHz crystal would be 2.45 x 0.0020 percent. At the frequency being monitored the percentage error would be 49/152.45 MHz x 0.01, or 0.000032 percent. This amount of error is not discernible on the meter.

P-8403-A

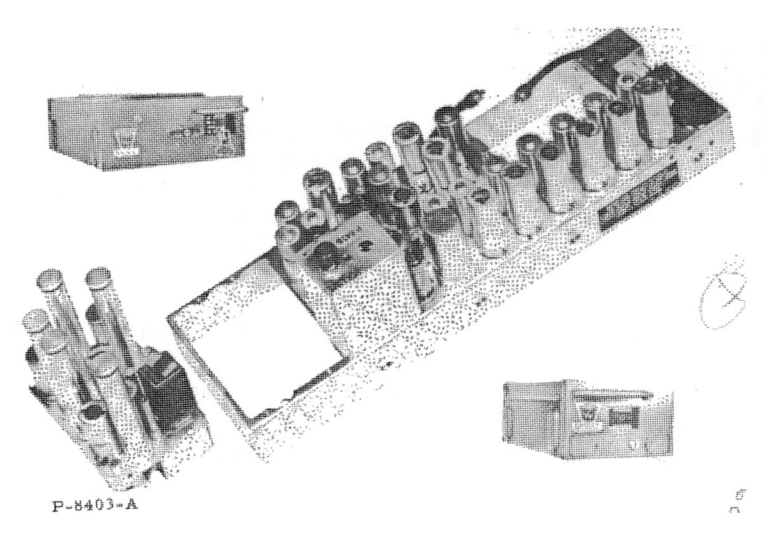

Figure 2

The "improvement factor" of possible percentage accuracy at the channel crystal frequency over the percentage accuracy at the carrier frequency is approximately the same ratio as the monitored carrier frequency over the channel crystal frequency. Hence, 0.0020 / 0.000032 equals 62.5 and 152.45 / 2.45 equals 62.2. This is another way of stating that the channel crystal is more than 62 times as good percentage-wise at the monitored frequency than at its fundamental frequency.

The improvement factor may be checked on any channel by the above method. It will always remain reasonably high; therefore, the possible error of the channel crystal frequency is negligible.

CONSTRUCTION

The front end deck of a Sensicon A receiver provides an ideal converter for the poor man's frequency meter. Figure 2 shows this assembly as a separate unit as well as in its original form installed in a Motorola Sensicon A receiver. The part numbers referred to in the modification procedures described here are those part numbers called out in the Motorola manual for the Sensicon A 150 MHz receiver. The procedure is quite simple, too. Here is all you do:

1. Replace R102 (2.2 meg) with 3.3 meg and ground low side.
2. Remove C104.

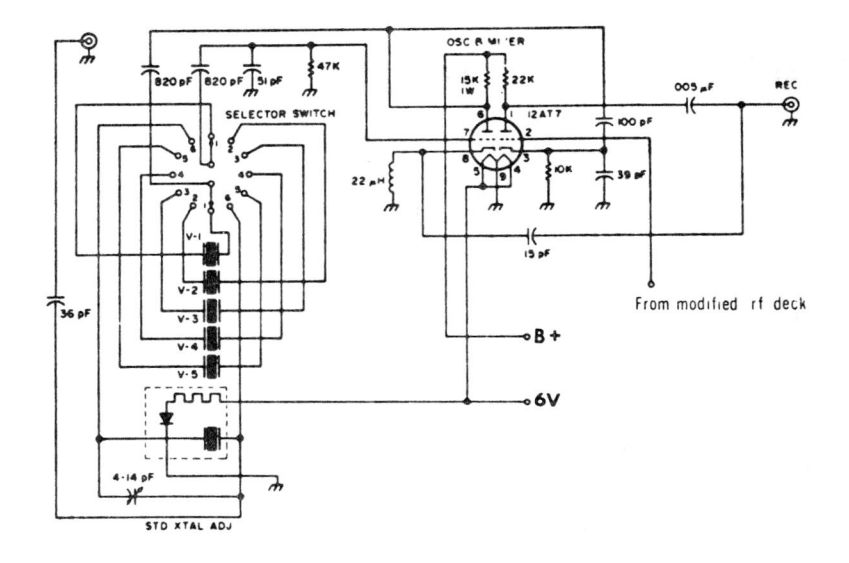

Figure 3

3. Replace R103 (33K) with 470K.
4. Replace L101 with a 100K resistor.
5. Remove R112 (3.9K), complete B-plus circuit.
6. Replace X102 with 9-pin socket (with shield).
7. Remove wire from pin 1 and connect to pin 2.
8. Connect 1 meg resistor from pin 2 to ground.
9. Wire 12AT7 socket for 6V filaments. (Connect 6V to pins 4 and 5, and ground pin 9.)

This completes the modification of the rf amplifier. To construct the oscillator / mixer, remove the balance of the circuitry on the deck with the exception of the crystal socket. Then build the circuit shown in Fig. 3 around the new X102. Be sure to use silver mica capacitors in the crystal circuits. The crystals themselves are Motorola SFMT-2 (R11, 5 MHz), and they may be obtained from Sentry or International.

The author gratefully acknowledges the technical assistance of Donald L. Milbury in preparing the information in this chapter.

Quickie CTS Decoder

by Louis LaBonte

Remote operation of repeaters is one phase of amateur radio that has been increasing tremendously in the past few years. Repeaters have been springing up in areas where, not too long ago, there was no activity at all—even of a direct nature. Inevitably, with the rise of repeaters comes the rise of interference—interference from areas often not even considered likely to be potential sources of trouble.

On my own remote base station (referred to locally as the "Voice of Auburn"), I experienced some very elusive interference on the control link that caused spurious keying of not only the UHF repeater output but the low-band base station as well. Inadvertent keying of a high-powered base station at a strategic location can cause a bit of dissention, as many—including me—can attest. The Voice of Auburn, during this interference phase of its career, drew its share of criticism from local 51 MHz stations attempting to communicate on the output channel.

Although the source of the interference was traced to something as innocent as a military altimeter, the fact that emissions from the remote transmitter hampered communications was sufficient to create on-channel friction. The answer to the interference problem, of course, was installation of CTS (continuous-tone squelched) decoding and encoding networks, referred to on the West Coast as PL, for private line. The big trouble here, however, was the fact that CTS equipment is hard to come by, difficult to build, and quite expensive when purchased new.

But it was a matter of either (1) designing relatively easy-to-build CTS circuitry, (2) spending beaucoup dollars for new stuff, or (3) staying off the air until the potential sources of interference were gone for good. The Scotsman's blood in me kept me from going the "buy new" route. And I wasn't about to stay off the air; for one thing, there will always be interference sources, and with the passage of time the sources are likely to increase rather than diminish. So I was left with the prospect of original design and construction.

The result of it all, I am happy to report, was positive. A simple decoder was developed using readily available (and cheap!) parts. This equipment was placed into operation at the repeater site, and the control point was outfitted with a matching encoder (much easier to scrounge than the decoder). After two years of operation, the system still works with all the security it was designed for.

The decoder unit is shown in the schematic. Its sensitivity is such that 1 to 3 kHz of CTS deviation on any 15 kHz system will provide reliable and troublefree operation with any signal capable of providing some degree of quieting into the repeater. The CTS frequency for which this decoder is designed can be anywhere between 80 and 200 Hz, as determined by the Vibrasponder reed (TU-333-L). The Vibrasponder may be obtained from radio service shops or other two-way service organizations. They are hard to come

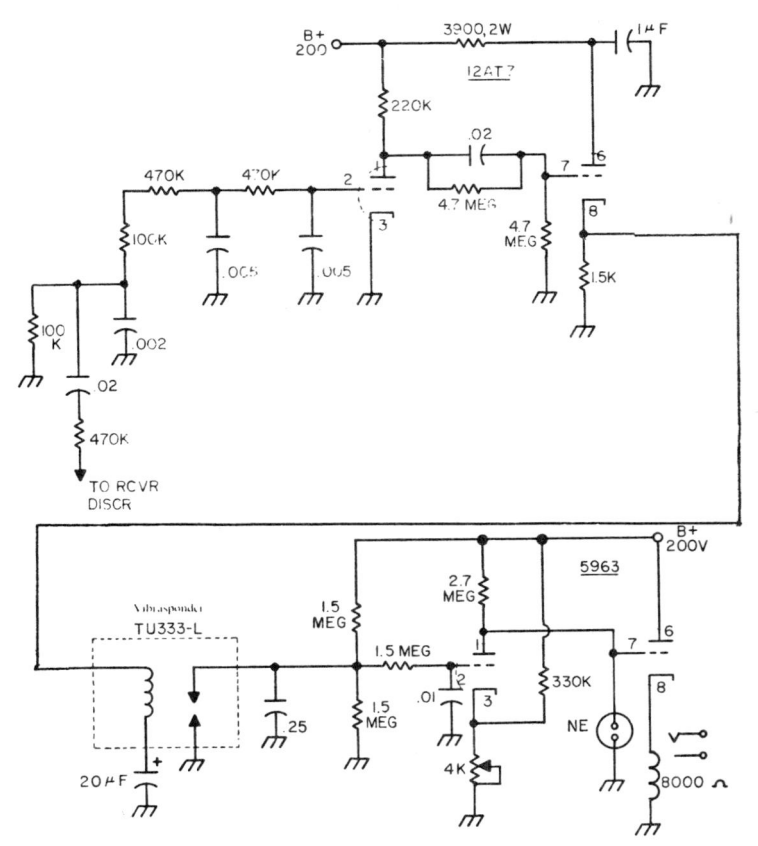

Figure 1

by. Motorola will not sell reeds without special authorization from their zone headquarters, I obtained them stating they were "for control of amateur repeaters...and not for commercial or public safety use."

Audio to drive the decoder is obtained from the discriminator of the receiver into which it is to be connected. The 12AT7 amplifies the signal to drive the reed. The output circuit (5963) drives a conventional 8-10K plate relay that may be used to key the transmitter directly or it may be connected in series with the repeater's carrier-operated relay for faster dropout. If there are additional functions to be triggered by the CTS decoder, it is a good idea to use the sensitive plate relay to drive a second relay constructed for heavy-duty applications.

The socket, a Motorola 9K832860, can be obtained from Motorola. No changes are necessary to change frequency other than replacing the reed Vibrasponder.

Frequencies around 100 Hz or lower are preferred as they do not pass through the receiver audio section easily. Usually a filter is not required to eliminate the tone if it is below 100 Hz and encoder level is set properly.

Frequency Synthesis: the Modern Way to Control Frequency
by Gilbert Boelke W2EUP

Frequency synthesis is the term used to describe the process of synthesizing (or "putting together") many frequencies from a small number of starting frequencies. In theory, any number of channels may be so generated from only one master oscillator, using the electronic techniques of adding, subtracting, multiplying, and dividing frequencies. In practice, the larger the number of channels the more worthwhile it is to go the synthesizer route.

A direct synthesizer uses such conventional techniques directly, filtering the undesired output products at each step in the process. This technique has the disadvantage that many high quality filters are required to reduce the undesired (spurious) output frequencies to the desired extent.

An indirect synthesizer uses a voltage-controlled oscillator to generate the output signal, electronically "steers" it to the correct frequency and "locks" it there. Its main advantage is that the output needs no filtering; it comes from an on-frequency oscillator. All spurious products are kept within the confines of the synthesizer loop (with any luck) and do not appear in the output.

Figures 1 and 2 illustrate two ways a synthesizer can be used. In Fig. 1, the synthesizer covers the full range of transmitter and receiver frequencies for a General Electric TPL unit. An extra x8 multiplier must be added to the receiver so that both receiver and transmitter multiply the synthesizer output by 24, to make channel spacing the same for both receiver and transmitter.

Although a synthesizer could be built as a crystal oscillator substitute for existing types of equipment, it can be more effectively exploited in a "start from scratch" design, as shown in Fig. 2. True FM can be produced by direct modulation of the synthesizer, eliminating the need for a phase modulator or frequency multiplication to build up the deviation level. Or, as a receiver local oscillator, a synthesizer can just as well be designed to deliver the oscillator injection frequency directly, eliminating the need for frequency multipliers.

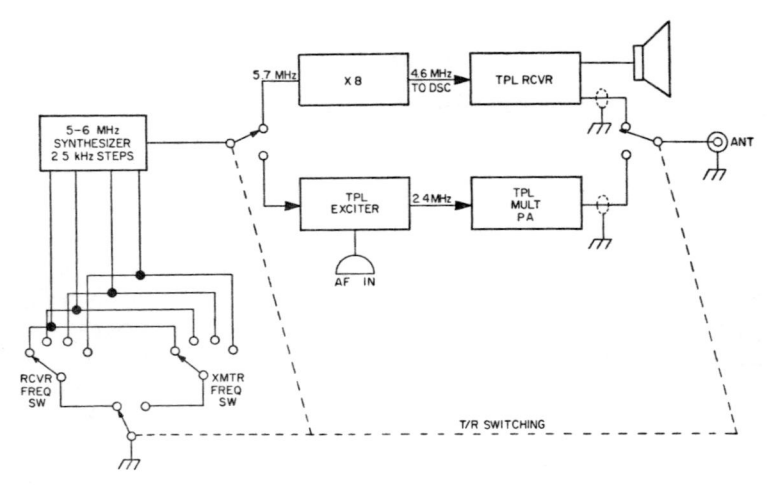

Fig. 1. How to use synthesizer for 60 kHz incremental switching in existing units.

Figures 3 and 4 show block diagrams representing the synthesizers used in Figs. 1 and 2. The synthesizer block diagram of Fig. 3 generates 2.5 kHz steps in the 6 MHz range. Used to drive existing transceivers, this arrangement produces 60 kHz steps in the two-meter band. If the range is extended to 5.7 MHz and the output multiplied by 24, the same synthesizer can be used for the receiver.

Addition of a mixer and a multiplier as in Fig. 4 makes it possible to generate 60 kHz steps directly in the two-meter band. A separate crystal oscillator is used to heterodyne the output signal down to a frequency suitable for division. By

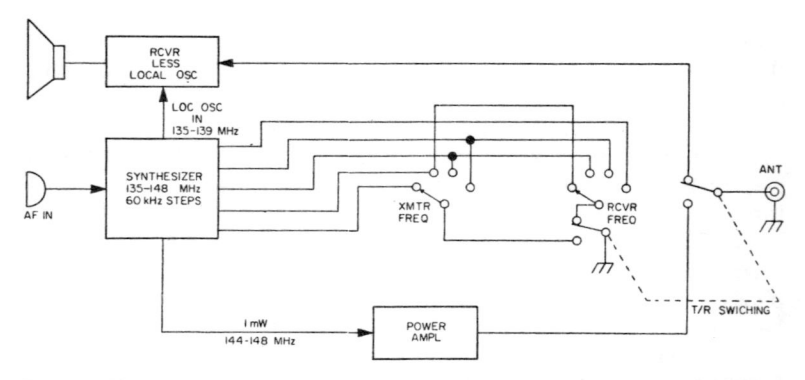

Fig. 2. Synthesis of local oscillator and transmitting frequencies eliminates frequency multipliers and greatly simplifies receiver design.

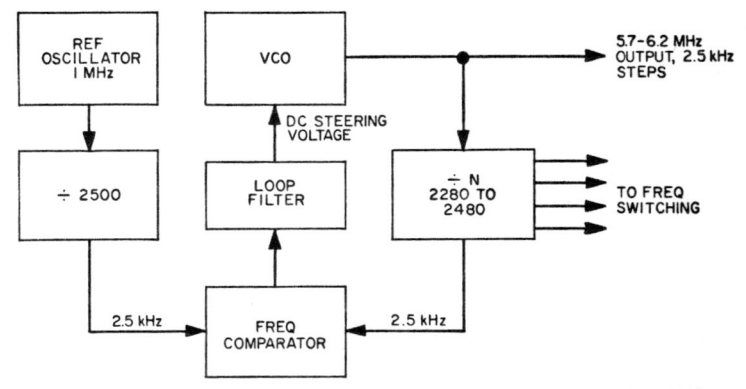

Fig. 3. Block diagram of synthesizer for use in existing transceivers (see also Fig. 1).

switching crystals—in this case to 141 MHz—the synthesizer output can be moved down 9 MHz to provide oscillator injection for a receiver having a 9 MHz i-f. These mixer injection frequencies could also be obtained by suitable means from the basic reference oscillator, rather than adding two more crystals.

How It Works

Consider the simplified synthesizer of Fig. 5. This example is for a 5-6 MHz output range in 10 kHz steps. A tunable VCO (voltage-controlled oscillator) is used to

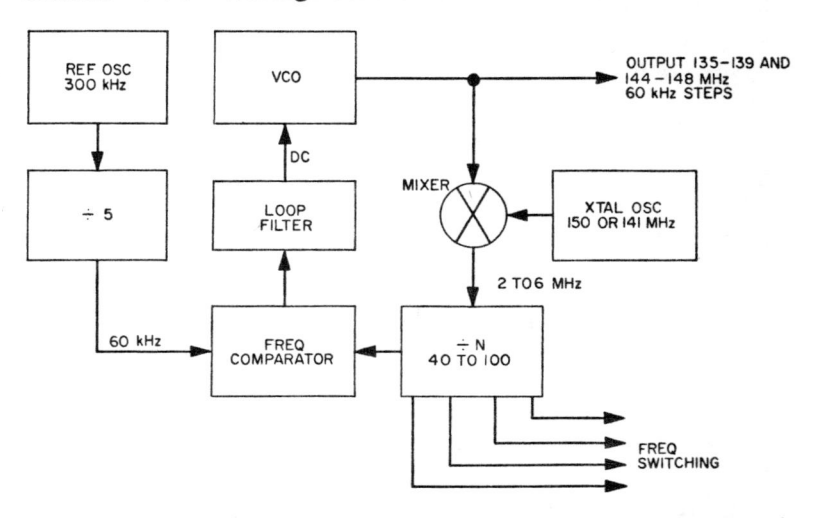

Fig. 4. Block diagram of synthesizer incorporating a mixer and multiplier for generating 60 kHz steps directly in two-meter band.

87

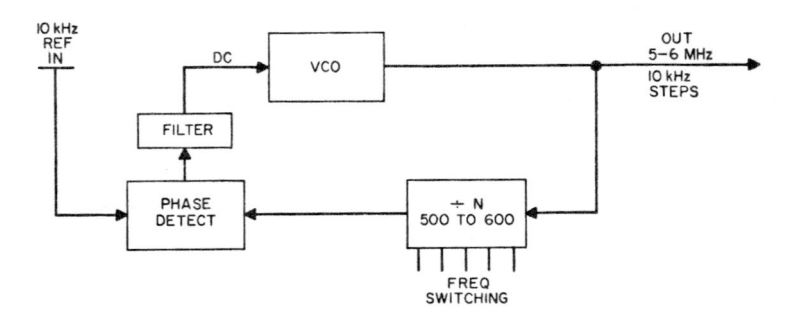

Fig. 5. Simplified synthesizer diagram.

generate the output signal; in doing so, of course, it must tune electronically from 5 to 6 MHz by varying the dc input voltage. The stability of such an oscillator doesn't even begin to match that of a crystal oscillator, but it does have the flexibility of operating on any channel in the desired range. So the remainder of the circuitry is devoted to detecting the VCO frequency, relating it to the desired frequency, and adjusting the oscillator electronically to it.

Spacing of the output steps is determined by the reference frequency. In this case, a 10 kHz reference means that the circuit can lock to any harmonic of 10 kHz such as 5.000, 5.010, 5.020, 5.030...etc. to 6.000 MHz. Direct multiplication could be used to get the same result, but it would be nearly impossible to eliminate undesired harmonics of the 1 kHz signal, even with the best of filters. So, instead of multiplying 10 kHz to, say, 5 MHz, start with the VCO at (or near) 5 MHz and divide it by 500 instead. This function is accomplished in the programmed divider (÷ n, or "divide-by-n"). Consisting of a chain of flip-flops, this circuit can be programmed to divide by any ratio between 500 and 600. When it divides the 5 MHz output of the VCO by 500, the result is 10 kHz.

The phase detector circuit compares the ÷ n output to the reference signal, delivering a dc output proportional to the phase difference between them. This dc level controls the VCO. When the VCO output drifts from exactly 5 MHz, the output of the ÷ n will drift in exact proportion to it since it always is 1/500th of the output signal. The phase detector sees this as a phase change and shifts its dc output to the phase detector immediately to steer it back to 5.000 MHz exactly. Since it compares phase instead of frequency, the frequency at the ÷ n output is never permitted to shift so much as one hertz either way, and the output of the VCO is held precisely to 5.00 MHz. Frequency accuracy depends only upon the stability of the reference frequency oscillator.

To change frequency to 5.760 MHz it is only necessary to change the divide ratio of the ÷n circuit by switching the programing inputs of the —n. Since this is a larger divide ratio than 500, the —n output will at first be below 10 kHz. The phase detector senses this shift as a phase change in the very first cycle following the shift and immediately starts action to correct it. When the correction is complete, the ÷n output is again on 10 kHz, but the VCO is on 576 x 10 kHz, or 5.760 MHz.

The synthesizer of Fig. 3 is only a little more complicated than the basic unit of Fig. 5. Since 2.5 kHz steps are desired, the reference frequency must be 2.5 kHz. Stable crystals of this frequency are not notably practical, so a 1 MHz master oscillator is used, divided down to 2.5 kHz by a series of flip-flops. A harmonic of the 1 MHz signal is conveniently adjusted to zero-beat with WWV, thus precisely aligning all channels to frequency. A higher divide ratio is necessary in the ÷n circuit due to the lower reference frequency, as shown.

The loop filter, necessary in all cases must remove all traces of the reference frequency at the output of the phase detector (in this case 2.5 kHz). A simple RC low-pass filter configuration is usually employed.

The next step in synthesizer development, shown in Fig. 4, has a VCO operating in the 135-148 MHz range, heterodyned to the 2-6 MHz range. This mixing process is necessary because present-day low-cost flip-flops can only divide to about 8-10 MHz in a programed divider circuit. A 60 kHz reference can be used to generate 60 kHz steps because the signal is not multiplied in the receiver and transmitter as in Fig. 3, thus simplifying both dividers greatly. Frequency stability in Fig. 4 depends mainly upon the accuracy of the 150 and 141 MHz oscillators. Frequency adjustment is necessary in all three oscillators with this scheme. Judicious selection of the frequency spacing, reference oscillator frequency, receiver i-f, and the mixer injection frequencies can result in a design that uses a single crystal.

The Phase-Locked Loop

Indirect frequency synthesizers are basically feedback systems, where phase error is detected and fed back as a correction signal. Such a closed loop is called a phase-locked loop. As a consequence, certain rules must be followed to keep the system stable, as in any feedback system.

The phase-locked loop must have a loop filter at the output of the phase detector to prevent rf from leaking into the VCO control signal (which should be as clean a dc signal as possible). If any rf or ripple appears at this point, it will

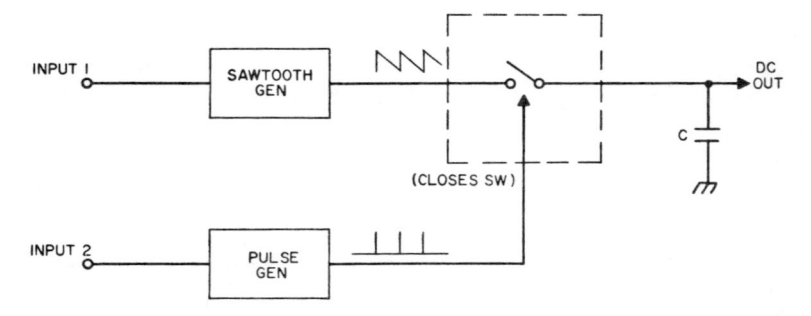

Fig. 6. Sample-and-hold phase detection.

frequency-modulate the VCO. If the ripple is deliberately applied as audio, a desired FM signal can be produced. Undesired high frequency components such as the reference frequency will produce spurious sidebands. Thus, it is the function of the loop filter to remove these undesired products. A second function, however, is to determine the phase—gain characteristics of the loop, which determine its response time, stability, and "capture range."

Capture range is the term applied to the maximum frequency difference the loop will tolerate between the VCO output and the desired output frequency and still lock up. Capture range is directly proportional to loop bandwidth. The higher this bandwidth, the higher the adjacent spurious levels—the lower it is, the closer the VCO must be before it will lock up to the desired frequency. A low loop bandwidth also takes longer to lock up. So a compromise is necessary. Despite all of these design criteria, the loop filter is usually a very simple circuit when used in conjunction with a good phase detector.

PHASE DETECTORS

Since the reference frequency of an indirect synthesizer is typically equal to the frequency spacing between channels, the phase detector also operates at this relatively low frequency. The best phase detector circuits are those which deliver the highest ratio of dc to inherent ac ripple. A flip-flop can be used as a digital phase detector, but its output is quite high in ripple content, and it requires a more sophisticated loop filter than other circuits. The best phase detectors in current use work on a sample and hold principle. One of the two input signals is converted to a sawtooth, the other to a narrow pulse. The former is called the ramp, the latter a sampling pulse. As shown in Fig. 6, the sampling pulse operates a series switch for brief intervals so that the value of the ramp voltage at that instant is transferred to a "holding" capacitor. If at the next sampling time there was no change in relative phase between the two signals, the output will not change from the first sample. If there is a difference, the output capacitor voltage shifts abruptly up or down to the new value. It can be seen that as long as the two signals are in phase lock the output ripple is (ideally) zero. With practical implementation it isn't zero, but it can be made extremely low with careful design.

False Locks

When the phase-locked loop is initially turned on or the frequency is changed, the VCO may be out of the capture range of the loop. Under this condition the synthesizer output is that of the free-running VCO: unstable and at an unknown frequency. The VCO must therefore be tuned to the near vicinity of the desired frequency before the loop will lock up. With most phase detectors a number of false lock conditions can occur. A false lock is present when the ÷ n and the reference frequencies are not equal, but the phase detector "thinks" they are and locks the VCO to the wrong place. A circuit which assures proper acquisition of the desired frequency is called a "search." It acts as an AFC-type control of the VCO by detecting frequency differences between the ÷ n output and the reference frequency, rather than phase differences. The search is normally turned off when the phase detector locks the loop. There are a number of ways in which a search can be implemented. Digital methods are compatible with the pulse-type signals available, and they offer simplicity of construction and freedom from adjustment. Best of all, they are nearly foolproof.

÷ n CIRCUITS

Except for the advent of low-cost integrated circuits, this part of an indirect synthesizer would probably be impractical to build. It consists of a chain of flip-flops whose function is to divide the VCO signal down to the reference frequency. The number of flip-flops depends upon the maximum divide ratio. Since each flip-flop can divide by a maximum of 2, two can divide by 4, three by 8, etc.; and n flip-flops can divide by 2^n. If the maximum divide ratio is 100, as in Fig. 4, it takes 7 flip-flops, which can divide up to $2^7 = 128$. Six would not be enough because they can divide by only $2^6 = 64$. In Fig. 3, a divide ratio of 2240 is needed; $2^{11} = 2048$, too low; $2^{12} = 4096$. Therefore, 12 flip-flops are needed.

The next problem is to find some way to change the ratio of the ÷ n by switching so that channels may be selected. Two common methods are used. To understand them, some of the properties of binary dividers must be known. First, the divider can be considered a counter since at each input pulse, or step, the flip-flops take on a unique combination of states. A useful analogy can be drawn to an automobile odometer (mileage indicator), which works in a similar way but counts in decades (powers of ten) instead of binary (powers of two). Including the tenths decade, a typical car odometer can count up to 999,999 tenths of miles, and the millionth step brings it back to all zeros. If a switch was provided to close only when all zeros are present, the switch would close once for every 1,000,000 input steps, thus dividing by one million.

One method of changing this ratio would be to reset the odometer to zero whenever it reached the desired count. For example, if a divide-by-567,000 count is desired, it could be accomplished by providing a resetting device which detects the 566,999 state, then resets all decades to zero on receipt of the 567,000th input. By programming the desired state when this occurs, the divide ratio can be any desired number.

Another way is to preset a number into the divider each time it recycles to zero. Achieving a ÷ 567,000 with this method requires that a 433,000 be inserted when the counter reaches a natural state of all zeros. The counter than counts from 433,000 to 1,000,000, where it is again preset. This count is 1,000,000 minus 433,000, or 567,000.

Presetting is the preferred method because it is usually easier to implement. Design of high-speed ÷ n circuits is full of subtleties which make it deceptively easy to design on paper, but another matter to make workable.

FREQUENCY DISPLAY

Up to about 12 channels, switching and display of the channel in use is a simple matter. For 30 or more channels it can become a problem, because even if a 30-position switch could be obtained, it would be considerably less convenient to find the channel you want than with only 12 channels.

For this reason, switches are usually arranged in a power sequence, such as a MHz, hundreds of kHz, and tens of kHz arrangement, extended to as many places as desired. If the n were left in its natural state, the switches would have to be set in a binary fashion, which could be awkward. However, the decade type of display is easily accomplished by designing the ÷ n circuit to work in decades instead of straight binary. It takes 4 flip-flops to produce a ÷ 10 section; cascading three such decades results in a capability of up to 1000 channels. If the channel kilohertz spacing is 1, 10, 100, etc., the frequency display is in familiar decimal numbers. Other schemes can be

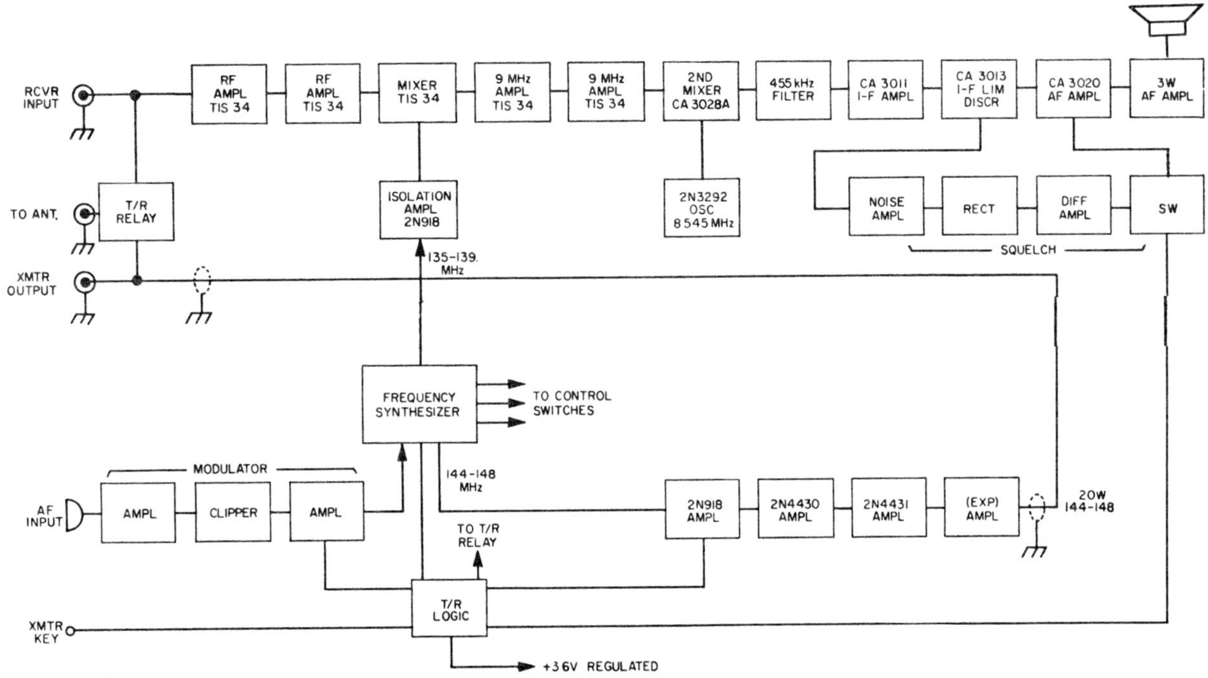

Fig. 7. Block diagram of an FM two-meter transceiver showing how frequency synthesis is incorporated.

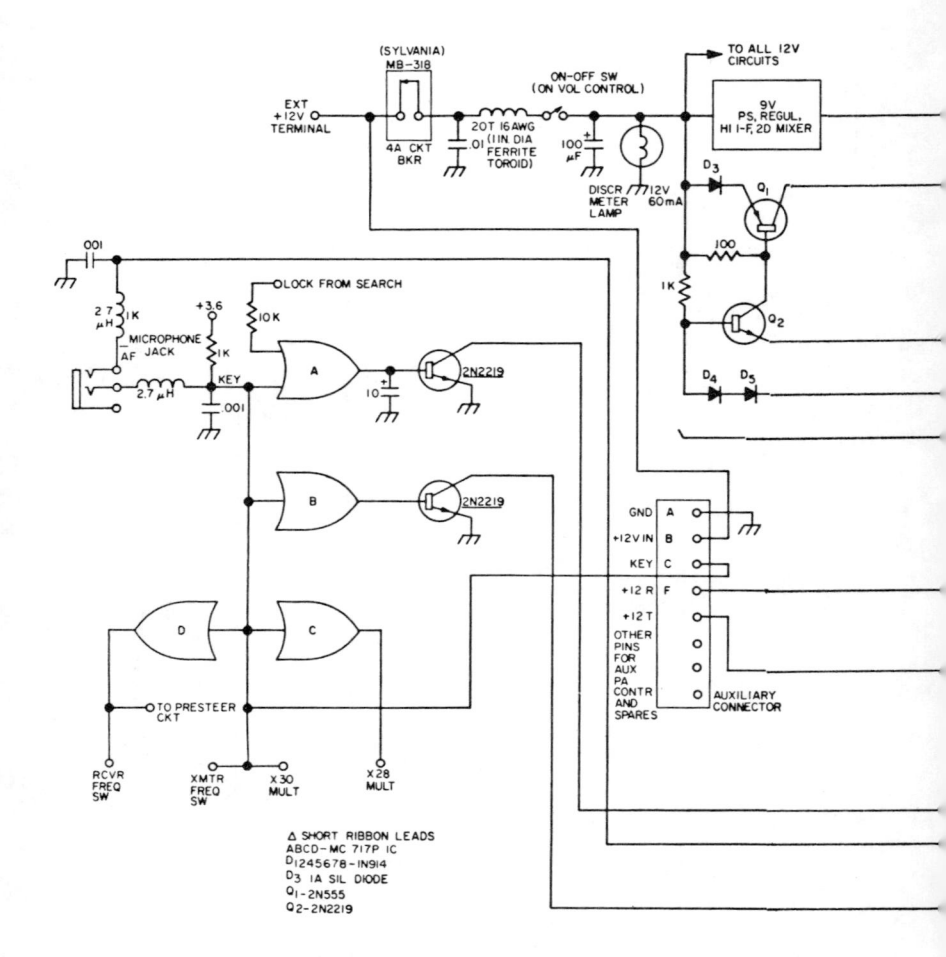

Fig. 8. Transmit and receive wiring for a typical FM transceiver.

worked out for different channel spacings, but the decimal method is the most convenient, since most of us think in terms of decimal numbers.

It should be kept in mind that the VCO output is the output of the synthesizer, and even though the loop keeps it exactly on frequency, it can't correct for audiofrequency variations (below the loop filter response). Even if it could, it would be necessary to slow it down for modulation purposes; otherwise the loop would attenuate the audio. Therefore, the VCO of a synthesizer is as sensitive to microphonics as a VFO is, and good VFO construction techniques should be used. VCO tuning range should just overlap each end of the desired output range, and its temperature drift should be kept low to maintain band coverage.

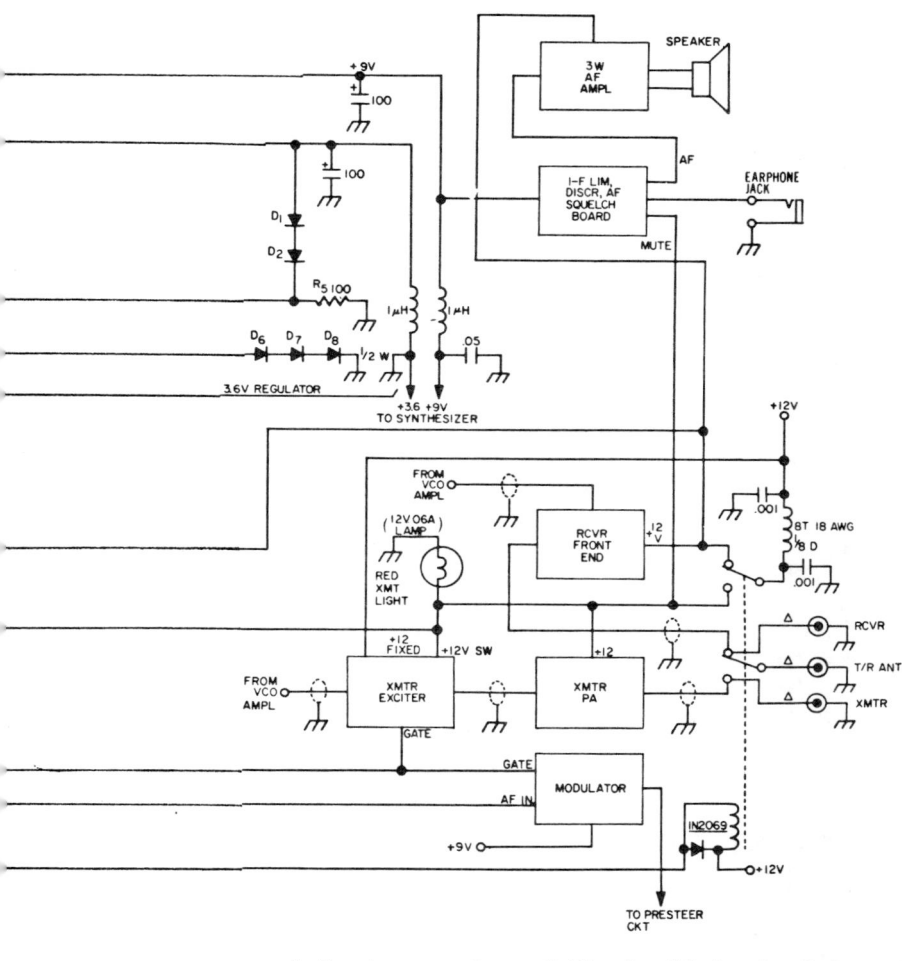

The following text describes the synthesizer used in a 10-kHz-step, 400-channel, 144-148 MHz transceiver utilizing a single 5 MHz crystal.

A PRACTICAL 10-kHz-STEP SYNTHESIZER FOR TWO METERS

Figure 7 shows the full transceiver block diagram. Output from the synthesizer is 1 mW in the frequency range of 144-148 MHz (transmitter), and 1 mW, 135-139 MHz, to the receiver first mixer. Modulation is applied directly to the synthesizer

Fig. 9. +9V regulator.

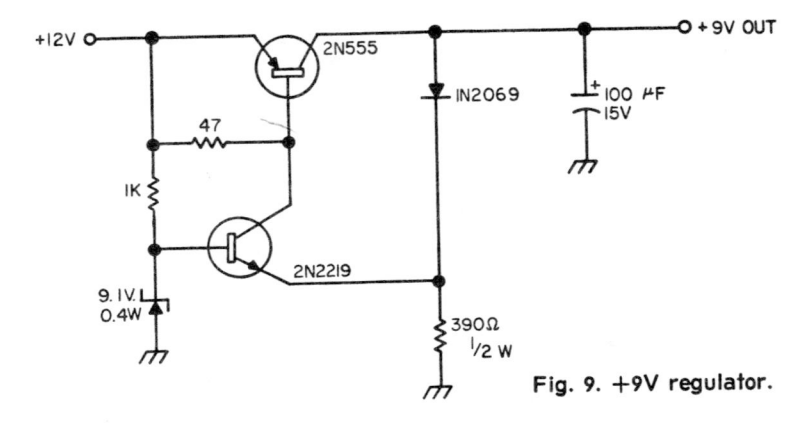

for the transmit function. The block labeled T / R logic is the circuit board that changes the synthesizer output range when the transmitter is keyed; it also serves to switch the receiver and transmitter and drive the antenna relay.

Figure 8 shows the transceiver transmit – receive circuits and the system wiring. Two power supply regulators are used; one generates +9V at 250 mA and the other +3.6V at 1A. Nine volts is the B+ level used in the receiver circuits, the transmitter modulator, and the synthesizer. The schematic for the +9V regulator is shown in Fig. 9.

The +3.6V supply is used to provide power for the digital integrated circuits, which are rated for a temperature range of +15 — +55°C (guaranteed performance). Instead of a zener diode, a series string of silicon diodes is used as a

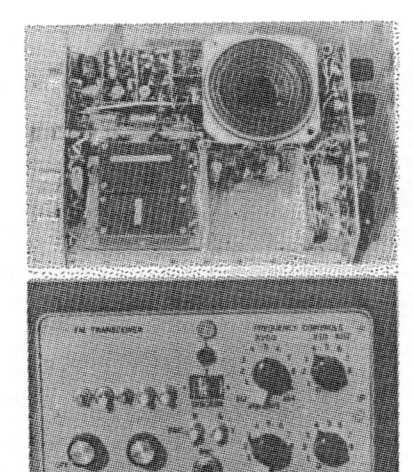

W2EUP'S homebrew two-meter FM transmitter—receiver looks unbelievably professional. The synthesizer controls are the four at the right of the panel. The top pair sets the receiver frequency, the bottom pair sets the transmitter. The basic frequencies of operation are selected by the integral four-position switches (under the X-100 knobs). The frequencies of operation on the pictured unit are 146.34 MHz for the transmitter (national repeater input frequency), and 146.94 MHz for the receiver (national repeater output frequency).

voltage reference for this regulator. This technique produces a temperature-programed supply that delivers over 4V when cold and less than 3.5V when hot. The logic works reliably over a very wide range of temperatures when operated in this manner.

Figure 10 shows the synthesizer block diagram. The single 5 MHz crystal oscillator drives, through an amplifier, a ÷ 500 fixed divider to obtain the 10 kHz reference frequency, and a dual-frequency multiplier section. When in the receive mode, a multiplication factor of X28 is used, producing a 140 MHz signal. When transmitting, a X30 multiplier supplies a 150 MHz signal to the mixer. The multipliers are selected by the transmit-receive (T/R) logic. When receiving, the VCO output range is 135-139 MHz (9 MHz below the corresponding transmit frequency). When mixed with the 140 MHz signal, the mixer output ranges from 5 to 1 MHz (140 — 135 = 5; 140 — 139 = 1) and the —n divide ratio is preset to divide by any number between 500 and 100, to deliver the 10 kHz output. Since 10 kHz is the reference frequency, the channel spacing is 10 kHz. In the transmit mode the mixer produces from 6 to 2 MHz

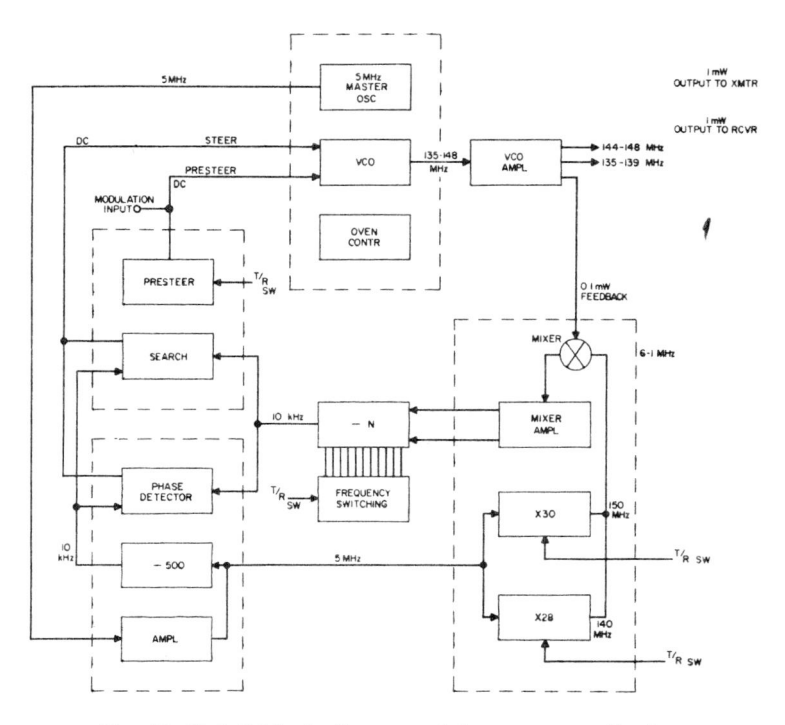

Fig. 10. Detail block diagram of frequency synthesizer.

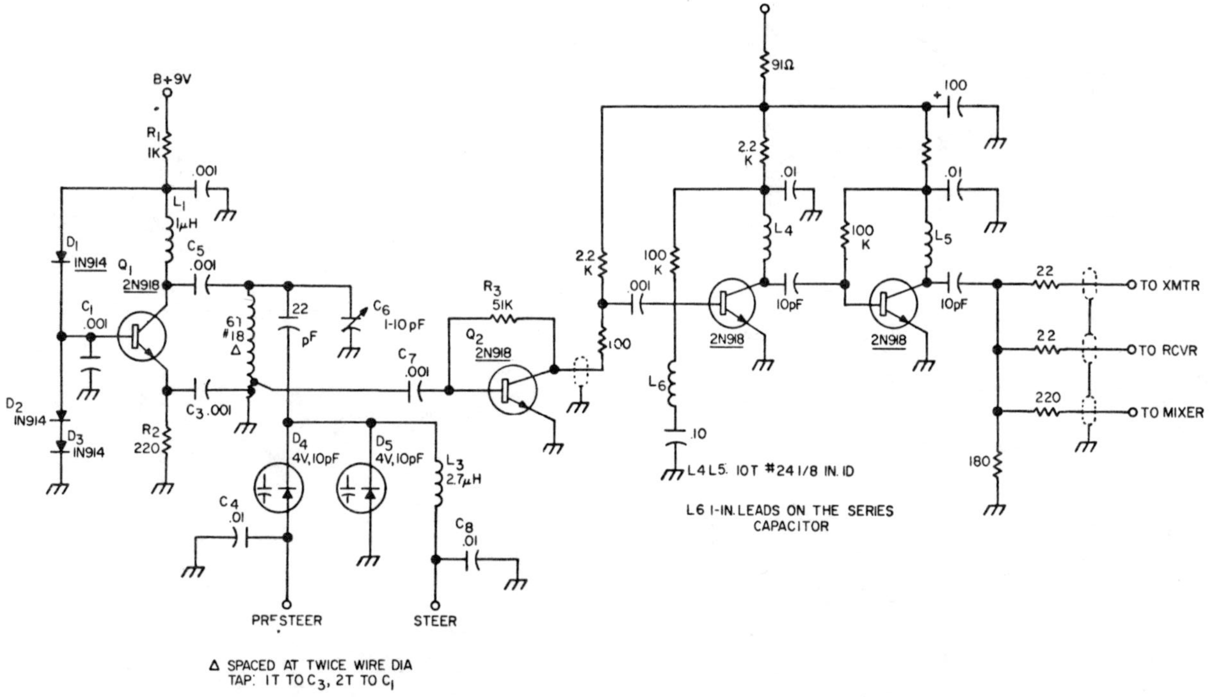

Fig. 11. Voltage-controlled oscillator and associated amplifier.

Detail of frequency selection switches show how "MHz" section connections are skewed 1 MHz offset on receive mode. At right an extra detent (switch clicking and locking mechanism) is reversed and mounted behind. A ⅛-inch shaft is attached and brought to panel through drilled-out shaft of front detent. Rear deck is X100 (hundreds of kilohertz, front two are MHz selection decks). Solderable magnet wire was used for connections to keep the wiring manageable.

(150 — 144 = 6; 150 — 148 = 2), and the ÷ n is preset to any ratio between 600 and 200.

The VCO and VCO amplifier are shown schematically in Fig. 11. The B+ to the VCO should be kept small to maintain maximum tuning range. As it turned out in the unit pictured, tuning range was no problem and had to be reduced by trimming the high end of the range with C6. The VCO is housed in a section of the oven assembly, which also contains the 5 MHz oscillator and an electronic temperature regulator. The VCO itself could have been mounted in a nonheated environment without stability problems because frequency drift with temperature turned out to be very low. However, in the author's transceiver, the oven insulation also acts as sound and vibration shielding (polyurethane foam), minimizing VCO microphonics.

Radiofrequency shielding is absolutely essential in the VCO. In the prototype unit, the VCO amplifier was mounted in a separate shielded box, but it was later combined with the VCO itself.

The steer input of the VCO comes from the frequency comparator (consisting of the phase detector and search) for electronic lock. An extra varactor (D4) is used to accomplish this by means of a voltage supplied from a presteer circuit (described later) and switched by the T/R circuits.

Three outputs are provided by the VCO amplifier: In addition to the receiver and transmitter outputs, there is a 0.1 mW signal fed to the synthesizer mixer for a loop feedback. The resistor network provides some degree of isolation between the outputs. Again, shielding of the VCO amplifier is a must.

Figure 12 shows the frequency multiplier section, the synthesizer mixer, and the amplifier used to square up the waveform to feed the ÷ n circuit. The multipliers are con-

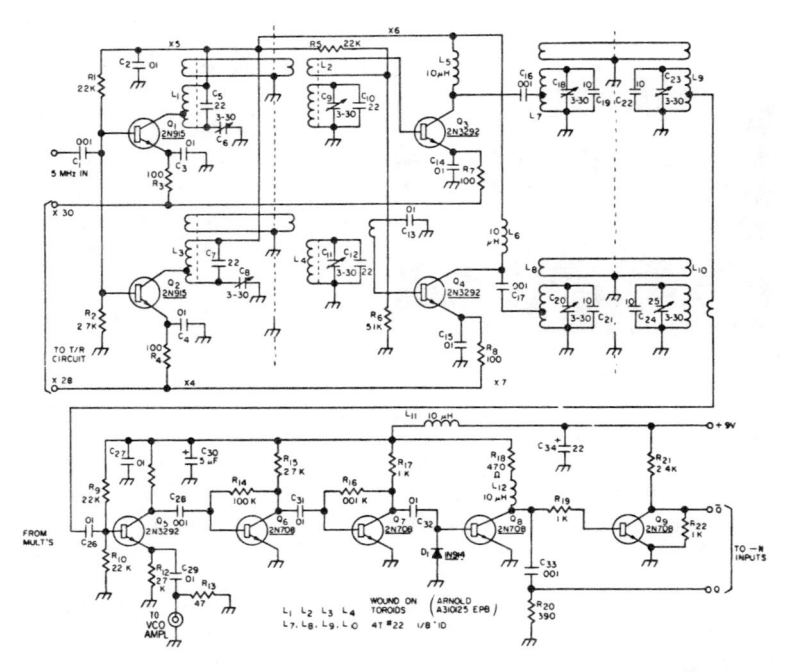

Fig. 12. Multiplier, mixer, squaring amplifier schematic diagram.

ventional and their outputs are connected in series to feed the mixer. Only one multiplier operates at a time, selected by the T/R circuits. Grounding the X28 or X30 lines selects the desired one.

Mixer output is amplified by several resistance-coupled stages. Transistor Q8 squares the waveform and Q9 acts as an inverter to provide the two out-of-phase signals needed to drive the ÷n circuit.

The ÷n schematic is shown in Fig. 13. Note that the divider is sectioned into decades I and II and a ÷8 section. The terminals shown along the bottom are the preset inputs. For maximum divide ratio (÷800) all of these terminals are biased to a positive 1-4 volts and thus no presets are inserted. Other divide ratios are chosen by selectively grounding (or opening) these terminals. For example, if A1 is grounded, it will divide by 799; if B1 is grounded, it divides by 798; C1 grounded yields ÷796, etc.

A table of presets is given in Table I, showing how the presets affect the divide ratio and the resulting frequencies of operation in this synthesizer. A zero indicates no preset; a "1" indicates a preset on terminals A1 through C3. Under the ÷n

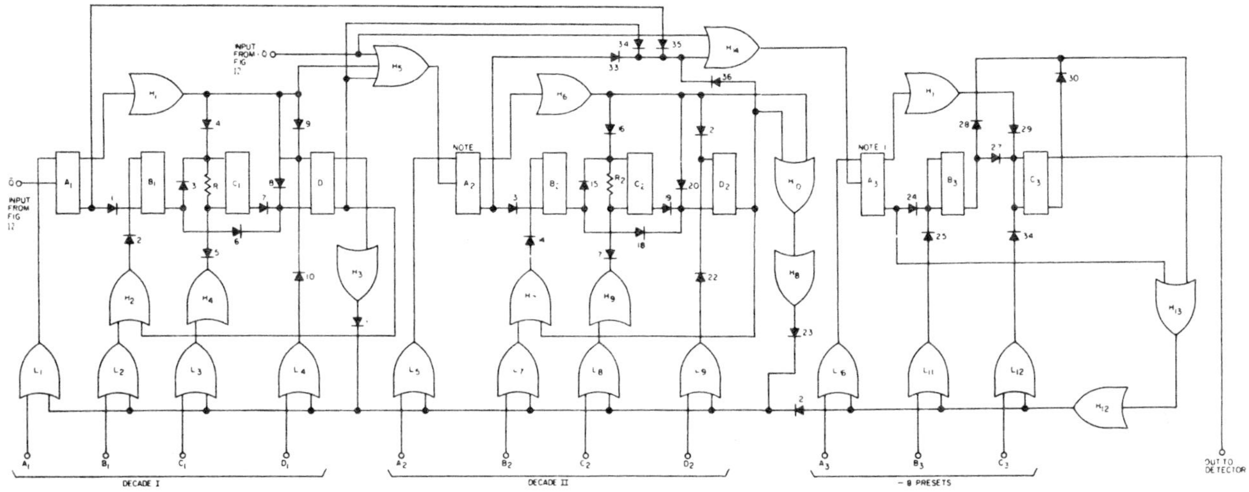

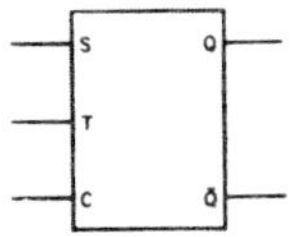

Fig. 13. ÷n circuit.

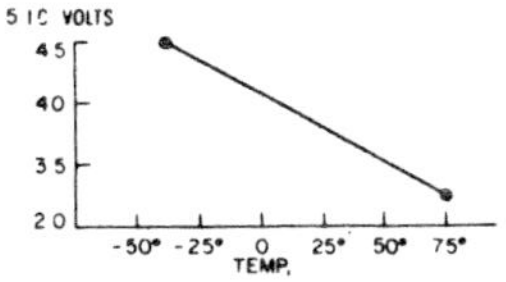

Table 1

Frequency		$\div n$ Ratio Reduction	A_1	B_1	C_1	D_1	A_2	B_2	C_2	D_2	A_3	B_3	C_3
0		0	0	0	0	0							
10		-1	1	0	0	0							
20		-2	0	1	0	0							
20		-2	1	1	0	0							
30		-3	0	0	1	0							
40		-4	1	0	1	0							
50		-5	0	1	1	0							
60		-6	1	1	1	0							
70	Steps,	-7	0	0	0	1							
80	kHz	-9	1	0	0	1							
0		0					0	0	0	0			
100		-10					1	0	0	0			
200		-20					0	1	0	0			
300		-30					1	1	0	0			
400		-40					0	0	1	0			
500		-50					1	0	1	0			
600		-60					0	1	1	0			
700		-70					1	1	1	0			
800		-80					0	0	0	1			
900		-90					1	0	0	1			
142		0									0	0	0
143		-100									1	0	0
144	XMT,	-200									0	1	0
145	MHz	-300									1	1	0
146		-400									0	0	1
147		-500									1	0	1
142		-100									1	0	0
143		-200									0	1	0
144	RCV,	-300									1	1	0
145	MHz	-400									0	0	1
146		-500									1	0	1
147		-600									0	1	1

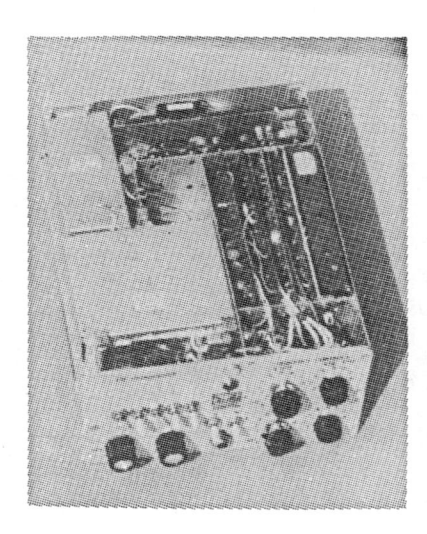

Top view of unit shows circuit board construction and "plug-in" accessibility concept of two-meter FM transmitter—receiver. The two rearmost cards are for receiver. Card next to i-f filter is i-f amplifier. The synthesizer section is shown at right. Left to right, the cards are: phase detect and divide-by-500; search; diode matrices; divide-by-n; mixer multiplier. The heavy white leads from switches to matrix board are insulating sleeves; each sleeve contains 8 — 12 leads of 32 AWG magnet wire. Slots and holes in shields are similarly guarded with insulation to prevent the possibility of shorts by chafing.

ratio heading is shown what happens to the normal ÷ 800 ratio as different combinations of presets are inserted. For example, a preset at A1 only (second line) shows a —1; this means that the ratio is reduced by a 1 and a 10, or 11; therefore, n = 789. Preset lines A1 to D1 thus switch tens of kilohertz, lines A2 to D2 go to the hundreds of kilohertz switch, and lines A3, B3, and C3 go to the megahertz switch.

A 10 MHz shift is accomplished in this design by changing the injection frequency into the synthesizer mixer. Since the receiver oscillator injection requirement is in the 135-139 MHz range (to produce a 9 MHz i-f), a 10 MHz shift is necessary to get the output into this range. As previously described, this function is accomplished by the T / R logic selecting the appropriate frequency multiplier. However, the MHz preset must also be shifted by 1 MHz to receive the same frequency as the transmitter is on, since only a 9 MHz offset is desired. Thus, as seen in the table, the receive presets are offset one place in the MHz column.

The ÷ n circuit shown is capable of operation to 10 MHz for all presets and represents the results of a hard brainstorming session. It should be reproducible and will work as it is shown, as long as the wiring is correct. Unless you understand its theory of operation completely it is recommended that you simply copy it carefully! Have someone else check every connection since troubleshooting is difficult. Space doesn't allow a complete explanation of its operation.

Frequency switching circuitry is shown in Fig. 14. Two complete sets of switches allow independent selection of receive and transmit frequencies. The diode matrices shown permit the use of standard 10-position rotary switches. Input voltage for the preset lines is provided through the 10K resistors. When one of the lines is grounded, a combination of presets is grounded through the diodes.

The arm of S1 is grounded in the transmit mode, and the arm of S2 is grounded in the receive mode. If S1 is in the A position, the transmit frequency is controlled by switch set A; if it is in the B position it transmits on the frequency on set B. The same is true for S2 on receive.

These switches provide great flexibility. With both switches in the A position, you transceive on frequency A. Similarly, with both in the B position you transceive on frequency B. Receive A and transmit B, receive B and transmit A are other combinations. It may sound complicated, but this system is very easy to use and a beginner can master it in seconds.

Once understanding of the method of controlling the synthesizer is complete, you can dream up all kinds of ways to

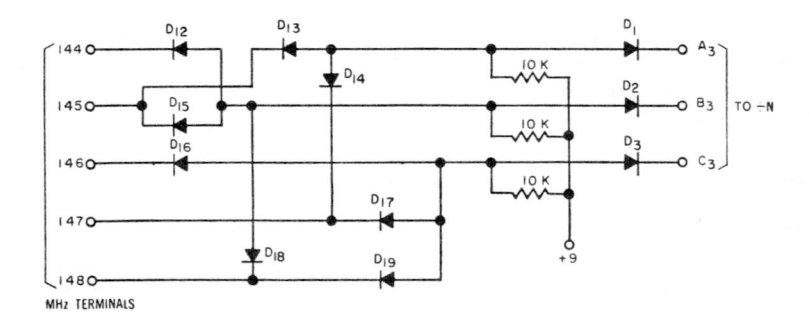

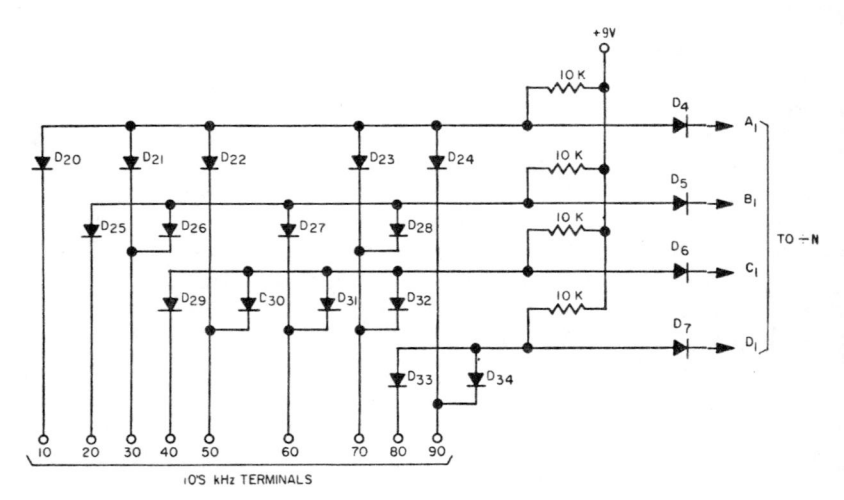

Fig. 14. Frequency-switching with diode matrices

Closeup photo shows that it is practicable to get all the components of the phase detect and divide-by-500 circuit onto a single small card.

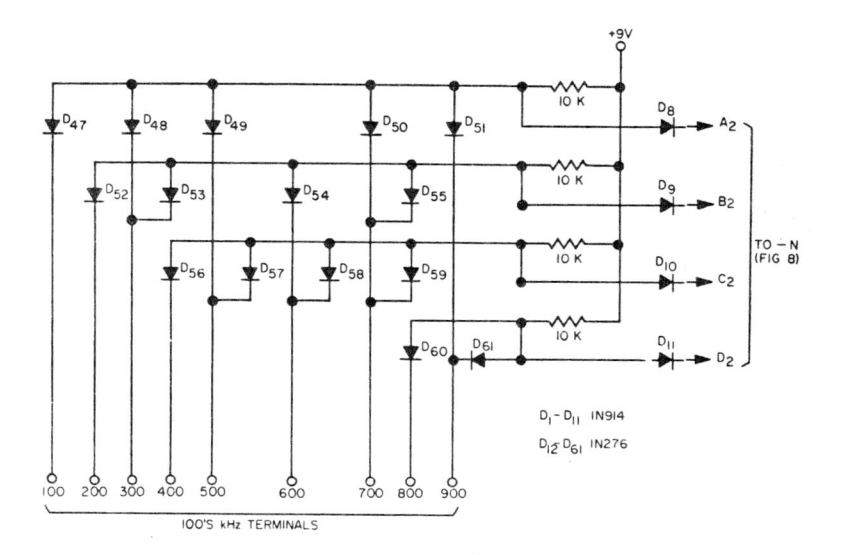

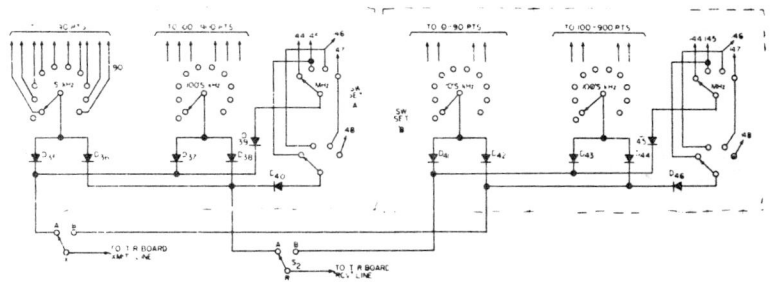

for rotary-switch utilization.

control it. For example, it is simple to instantly select a preset channel, such as 146.94 (the national FM repeater output frequency) by throwing a toggle switch. This addition can be completely independent of other switch positions. And remote switching is easy because all control lines to the ÷ n circuit handle only dc and simply are grounded in different combinations.

Figure 15 shows the master oscillator. It uses a field-effect transistor and has an automatic gain control arrangement for high stability. It is completely shielded and oven-controlled (Fig. 16). Asterisked components are thermally mounted to the oven box; the thermistor senses the oven temperature, and the other asterisked components deliver heat to it. In the original unit, the temperature control pot is mounted outside of the oven.

105

Figure 15

MASTER OSCILLATOR

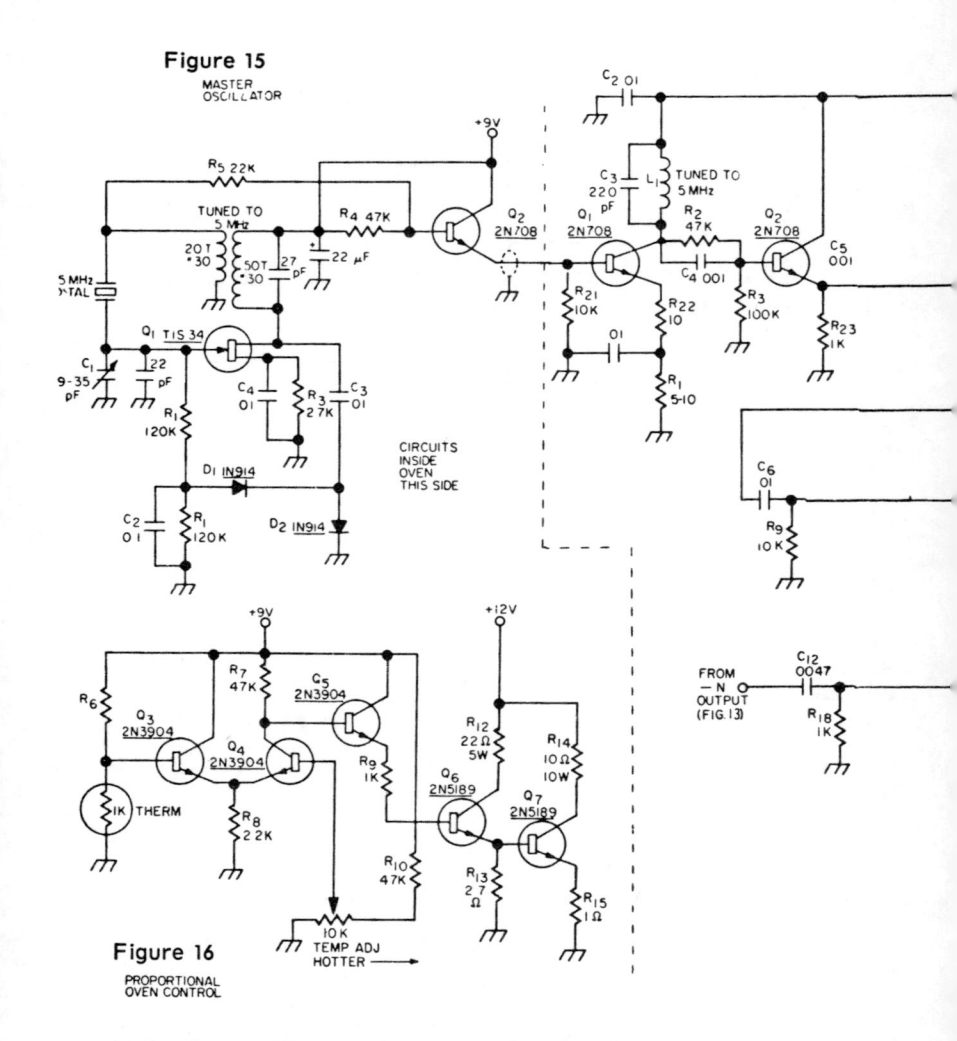

Figure 16

PROPORTIONAL OVEN CONTROL

Referring to Fig. 17, the 5 MHz low-level signal from the master oscillator is amplified in Q1 and Q2. Output from Q2 goes to the multiplier section and to squaring amplifier Q3, which drives the ÷500 circuit. Two outputs are used from the last flip-flop (I) 180 degrees out of phase. One goes to the search circuit, the other to the phase detector, below. The 10 kHz square wave is converted to a spike by C6-R9, as seen in waveforms a and b in Fig. 18. Each positive spike turns Q5 on momentarily, charging C7 to +9V. Between spikes (waveform c), C7 discharges through R15, producing a sawtooth. (A linear sawtooth could be used instead of the nonlinear one used

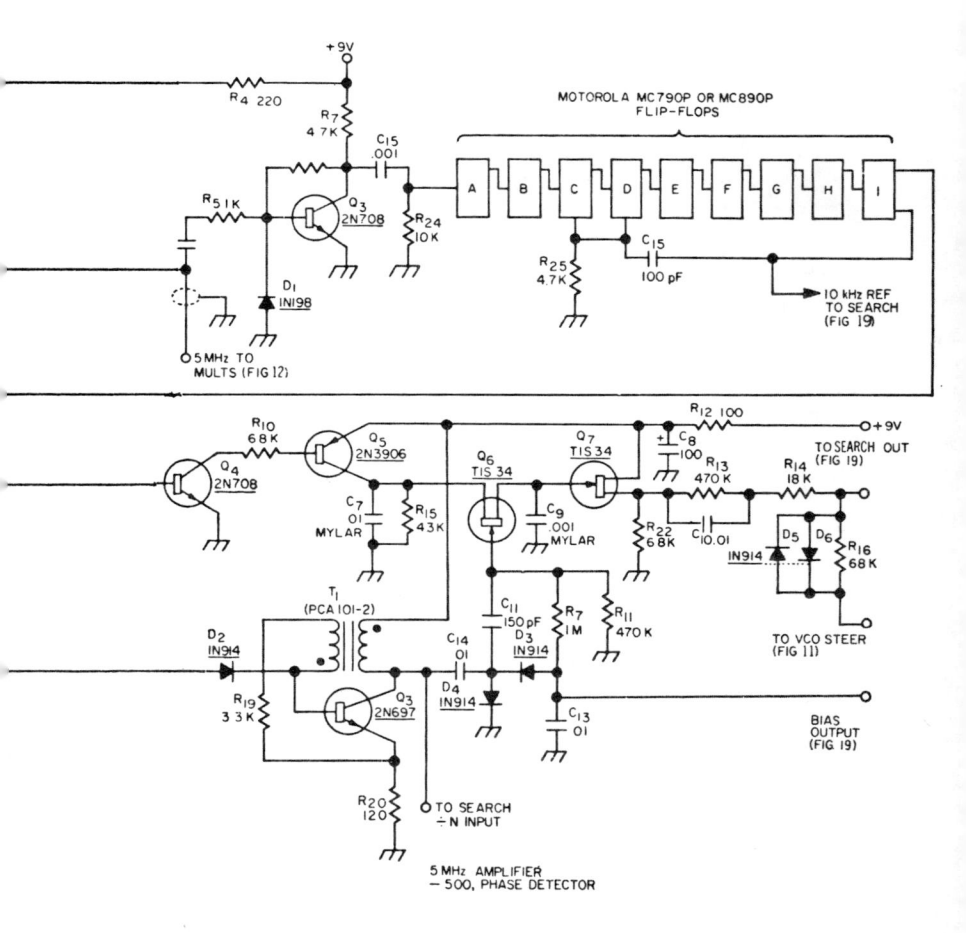

Fig. 17. Phase detector waveforms.

here, but the nonlinear waveform is actually beneficial in this system and is easier to generate.)

Sampling pulses are produced from the ÷ n output in a blocking oscillator (Q3) and fed to the gate of Q6 (waveform d). This pulse turns Q6 on briefly, charging or discharging C9 to the value of voltage on C7 at that instant. Capacitor C9 can only charge or discharge through Q6, so it holds that value of voltage until the next sampling pulse. Different ÷ n outputs and the resulting voltages across C9 are shown in Fig. 18, d through i. Transistor Q7 is a source follower which drives the following circuitry at a low-impedance level, while main-

Closeup photo shows the rearch board at left, the divide-by-n card and, at right, the mixer—multiplier assembly. The wall shield between those boards still in place are made from a conducting material deposited onto the fibrous board material; the close spacing increases the possibility of card-to-wall shorting, so a Mylar insulation sheet was attached over the surface of each of the shield walls.

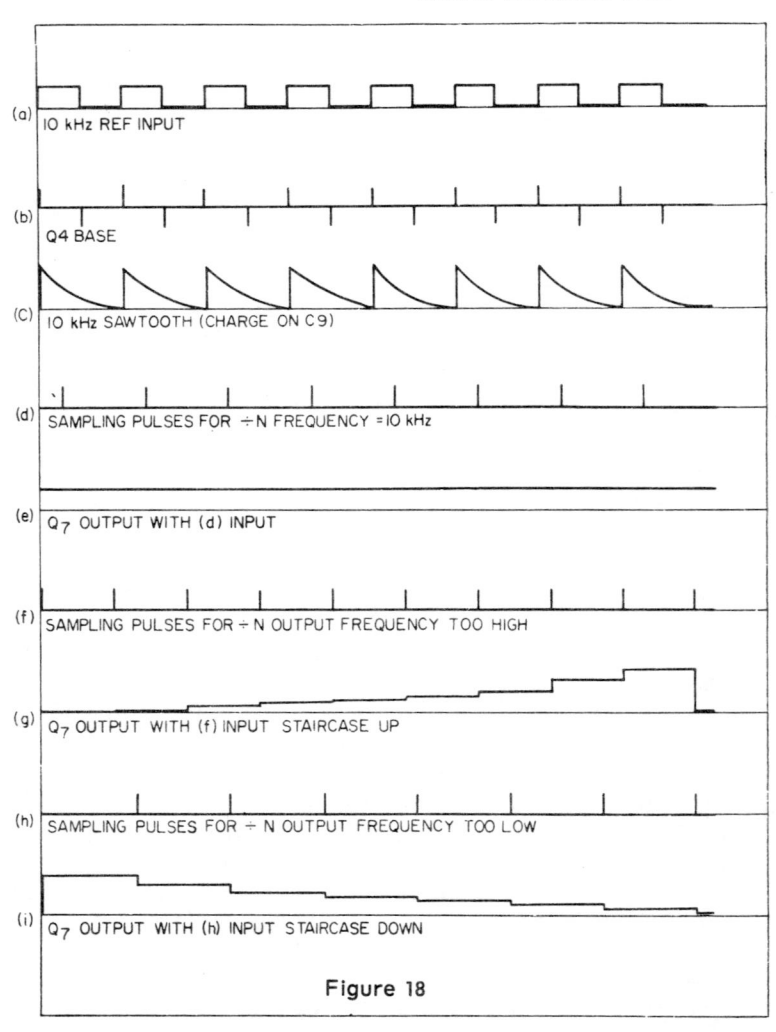

(a) 10 kHz REF INPUT

(b) Q4 BASE

(c) 10 kHz SAWTOOTH (CHARGE ON C9)

(d) SAMPLING PULSES FOR ÷N FREQUENCY = 10 kHz

(e) Q_7 OUTPUT WITH (d) INPUT

(f) SAMPLING PULSES FOR ÷N OUTPUT FREQUENCY TOO HIGH

(g) Q_7 OUTPUT WITH (f) INPUT STAIRCASE UP

(h) SAMPLING PULSES FOR ÷N OUTPUT FREQUENCY TOO LOW

(i) Q_7 OUTPUT WITH (h) INPUT STAIRCASE DOWN

Figure 18

taining a near-infinite load on C9. The loop filter consists of R13, C10, R14, R16, diodes D5 and D6, and the rf bypasses at the VCO. The diodes effectively short out R16 for sudden large shifts in phase-detector output to speed the lockup process. For small changes, as seen when the loop is in lock, they have no effect.

A bias voltage is developed from the high-amplitude blocking oscillator output with D3 and D4. This bias is used to hold Q6 off between sampling pulses and to bias the varactor presteer input on the VCO.

Figure 19 shows the search circuit. It operates from the same two frequency inputs that the phase detector uses, except that its purpose is to detect frequency instead of phase differences, and to coarse-tune the VCO to the desired frequency, where the phase detector takes over. It accomplishes this by checking for pulse interlace; that is, to see that for every pulse received on one input there is only one pulse received at the other input. Obviously, if two pulses occur from one input during which time no pulse from the other input is received, the two pulse trains can't be of the same frequency.

Figure 20 shows waveforms for the locked condition (a, b, c) where the comparator, consisting of gates BCGEHF, etc., does not produce any output pulses; and where the ÷n output is too high in frequency (d, e, f). When this case exists, gate H delivers pulses. When the ÷n output is too low in frequency, gate F puts out a series of pulses. When pulses come from H, Q6 and Q3 are pulsed on, producing a stepwise increase in voltage across C9. This voltage is summed with the voltage from the phase detector. As it rises, the VCO is tuned higher in frequency, which decreases the output frequency from the synthesizer mixer, decreasing the ÷n output frequency as desired. A correction in the opposite direction is accomplished by pulsing Q4 from gate F, decreasing the C9 charge in steps.

Gates I, J, and D do two things: they gate the transmitter off while the loop is searching (so you don't search while on the air) and they drive the out-of-lock indicator light (which tells you when something is wrong). The indicator normally flashes briefly between receive and transmit. If anything goes wrong in the synthesizer the light is almost sure to indicate it.

Transistor Q8 is used as a presteering gate. Controlled by the T/R circuit, it is biased on in the receive mode, placing a positive voltage on the presteer input of the VCO. This voltage, adjusted by R16, reduces the voltage across varactor D4, increasing its capacitance and shifting the VCO tuning range down. When Q8 is off (in the transmit mode) the bias is

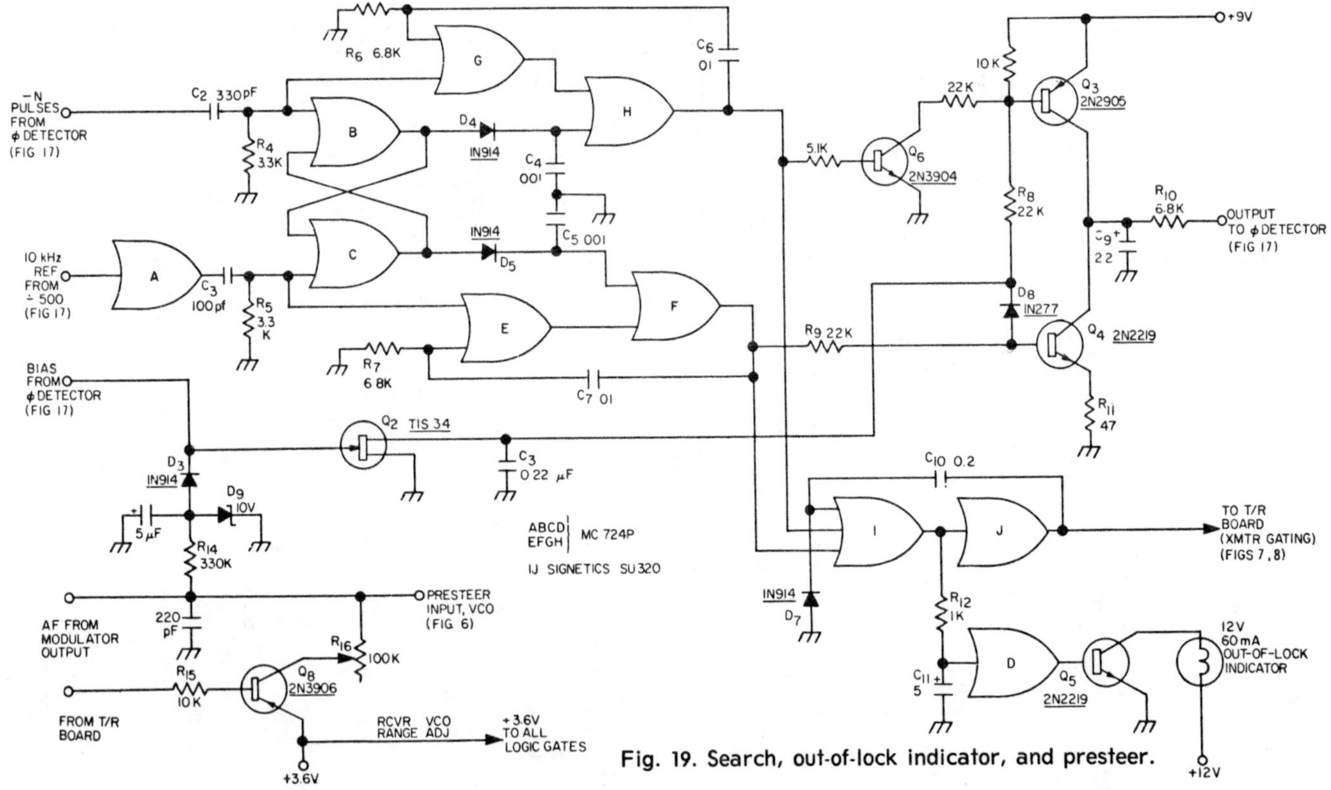

Fig. 19. Search, out-of-lock indicator, and presteer.

(a) |||||||||||
DIFFERENTIATED 10 kHz REFERENCE INPUT

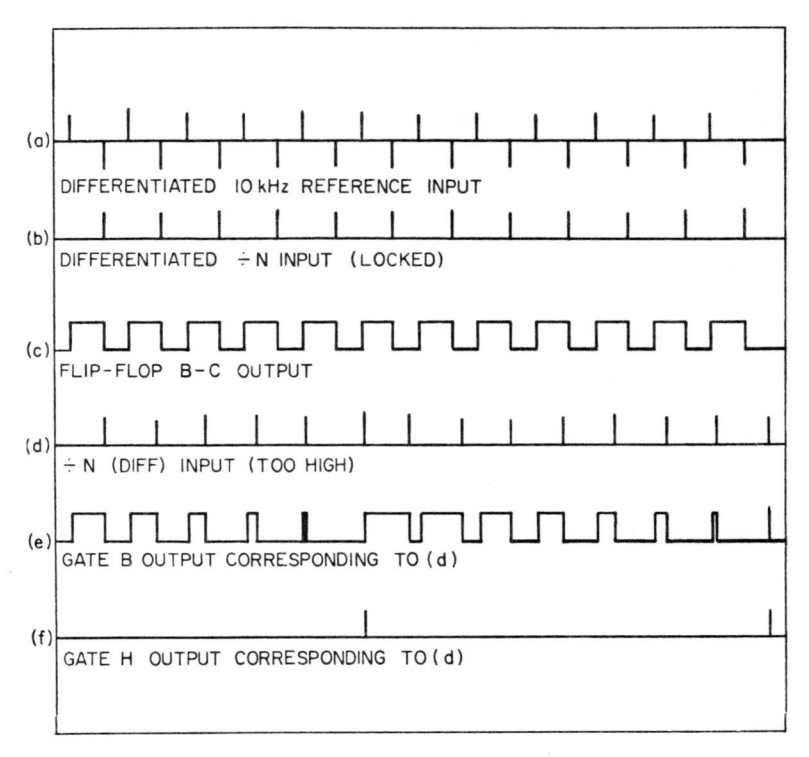

(b) |||||||||||
DIFFERENTIATED ÷ N INPUT (LOCKED)

(c) ⎍⎍⎍⎍⎍⎍⎍⎍⎍⎍
FLIP-FLOP B-C OUTPUT

(d) |||||||||||||
÷ N (DIFF) INPUT (TOO HIGH)

(e) ⎍⎍⎍⎍⎍⎍⎍⎍⎍⎍
GATE B OUTPUT CORRESPONDING TO (d)

(f) | |
GATE H OUTPUT CORRESPONDING TO (d)

Fig. 20. Search waveforms.

allowed to swing to —10 volts, which reduces the capacitance of D4 to a minimum. Diode D9 regulates this bias level. Modulation is ac-coupled to the presteer input instead of the steer input so that it does not interfere with the operation of the loop. The 220 pF capacitor is an rf bypass. Modulation input impedance is 330K, and very little voltage swing is needed for 15 kHz deviation. Linearity for this level of deviation is excellent and hi-fi audio is possible.

Transistor Q2 is used to prevent a possible hangup condition of the loop, where the VCO gets tuned so low that the frequency supplied to the ÷ n is beyond its counting capability. The ÷ n would then start counting erratically, delivering too few pulses instead of too many, due to skipping pulses. The search circuit interprets this to mean that the ÷ n output is too low instead of too high, so it steers the VCO in the wrong direction, perpetuating the situation. Luckily, the bias supply in the phase detector happens to depend upon a continuous supply of ÷ n pulses, so that when this hangup condition occurs, it can be detected by a large drop in bias voltage.

The oven assembly should be thermally isolated to the greatest extent possible. In the unit pictured, the crystal oven circuit is isolated from the remainder of the circuitry by means of a thick styrofoam surround. Shown in the oven are: the crystal (at left), voltage-controlled oscillator (in corner compartment), and 5 MHz oscillator with oven control circuit.

Transistor Q2 is normally biased off by this supply, but when the loop hangs up, bias collapses, so Q2 turns on and turns search transistor Q3 on in good Rube Goldberg fashion. Transistor Q3 charges C9 to maximum voltage, sweeping the VCO to the high end of its range, where normal lockup can take place. The entire process takes place in a few milliseconds. C3 causes a delay to make sure that Q3 turns on completely.

Shielding

Most of the circuits are susceptible to the high-level rf fields typically generated by an adjacent transmitter. The ÷ n circuit is an efficient hash generator to nearby receivers at nearly any frequency. It is therefore important that most of the synthesizer circuits be shielded from the outside world as well as from each other. All leads should be filtered and bypassed with RC or LC circuits. Extra B+ bypasses in the system will be found helpful in various spots. The most insidious form of system trouble is when complex circuits interact in ways not anticipated; so make sure the circuits function as they should separately, and then combine them in sections. Test everything for proper function before making any attempt to close the loop. When the loop is out of lock, everything jumps at once, and it is truly enough to make a grown man cry. Usually, the only hope is to open the loop and check individual circuits. With experience you can read the signs and troubleshooting becomes just as easy as robbing Fort Knox.

Performance

With a good master oscillator you can get counterlike frequency accuracy on all channels. The author's unit is accurate to better than plus or minus 20 Hz at two meters with a 20 minute warmup. Even without a warmup period, it is considerably better than most crystal-controlled rigs after stabilization.

Transistor Preamp for Hi-Fi Audio

by Dick Thomas

Audio quality is becoming increasingly important for FM operation, and for some very good reasons. The advent of repeaters that is now upon us almost guarantees that the average operator will be communicating with the aid of a repeater at least part of the time. But the repeater, with all its range-extending capabilities, poses special problems in terms of signal intelligibility, because the repeater can't help but introduce some distortion to a signal, and it compounds problems of less-than-perfect audio.

A carbon microphone, though considered the workhorse of communications, does not lend itself well to high-fidelity audio transmission. It reproduces a narrow range of voice frequencies well enough, but it attenuates the highs and lows and adds such a high degree of coloration to the modulating voice signal that unintelligibility often results when the signal is regenerated through a repeater. And when the repeater's audio system is less than the ultimate, or when multiple repeaters are used, the carbon microphone can no longer be considered as a candidate.

One excellent way to get around the problems of low-fidelity audio transmission is to use a dynamic microphone. Typically, its reproduction range extends far below and above that of the carbon mike, and the coloration is nowhere near as severe. One nice thing about converting to a dynamic microphone is that no modifications are required to the transmitter audio circuits. The dc drive voltage once used for the carbon-mike excitation voltage, can be applied without change to a tiny transistor amplifier circuit which can be readily incorporated into practically any standard microphone case.

The amplifier circuit shown in Fig. 1, for example, will fit beautifully inside the case of an Electro-Voice Model 602 differential dynamic microphone housing. The amplifier has a gain of about 35 dB, which serves to bring the microphone signal up to the normal output of the carbon microphone.

Once the amplifier is built and a dynamic microphone is added, the unit will plug directly into the mike connector of

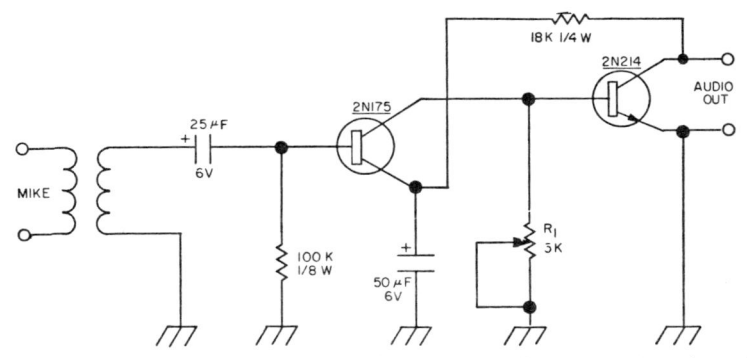

Fig. 1. Transistor preamplifier can be used with dynamic microphone to replace carbon microphone for improving fidelity of transmitted audio. Circuit can be nestled into most conventional microphone housings.

any standard FM transmitter. I have used the circuit successfully with Motorola, GE, RCA, and Narco. The quality is always high and audio is always in abundance.

Most any audio transistors (one NPN and one PNP) will work. The NPN should have a dissipation rating of at least 125 mW.

There is nothing critical in the selection of the microphone, either; most any dynamic, ribbon, controlled reluctance, or magnetic unit will prove ideal. The circuit will work equally well whether the mike element is high- or low-impedance, though the high-impedance types will tend to give a few decibels more gain.

To get optimum performance from your unit, vary R1 for best audio quality and volume. The total current measured on the dc line should not exceed 9 mA. For compactness, R1 may be replaced with a fixed resistor rated at one-eighth watt.

A test circuit may be set up using a 6-volt battery and a 500-ohm transformer (to simulate the input transformer in the transmitter itself), as shown in Fig. 2. Connect a pair of high-impedance headphones to the transformer secondary so you can monitor the amplified audio. The point of best quality will probably be when R1 is between 3 and 4K, and when the total battery current is between 3 and 6 mA.

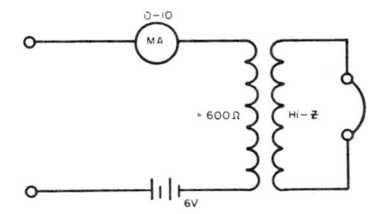

Fig. 2. Test circuit can be built up to optimize the quality of the preamplifier by experimenting with component values and monitoring with high-impedance headphones.

EQUIPMENT CONVERSION

AM to FM ...
in 10 Minutes!
by Gardner Harris

Are you one of the small percentage of readers who has never transmitted on FM simply because you had no FM equipment or no way of frequency-modulating your existing gear? Now there can be no excuse, for here are descriptions of two simple modification circuits, one of which will almost instantaneously put your vfo-operated transmitter on FM. The other is a bit more complex, but will do an equally good job on your crystal-controlled job.

Both modulators should be driven with a high output, high impedance crystal or ceramic microphone. No audio amplification other than that provided by the circuits themselves will be necessary.

The vfo varactor modulator, as simple as it is, seems to be a little-known but most effective method of producing frequency modulation; it utilizes the minute voltage generated by the piezoelectric action of the microphone element to cause the varactor to shift the vfo frequency at an audio rate, thus producing true FM (as opposed to phase modulation, which is found in most commercial gear). Although the actual shift is only a few hundred hertz, by the time the vfo frequency is multiplied 18 times (assuming an 8 MHz oscillator frequency) the total deviation will be more than enough to suffice for good communication.

The vfo varactor modulator shown in Fig. 1 may be assembled on a standard three-terminal phenolic tiestrip and tucked into a corner of most any vfo chassis. If all parts are purchased new, the total cost will be less than two dollars.

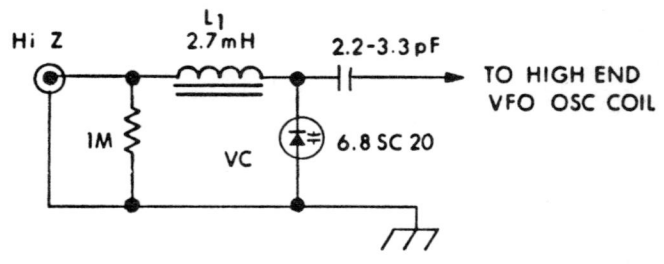

Fig. 1. Varactor frequency-modulator for vfo-equipped AM rigs.

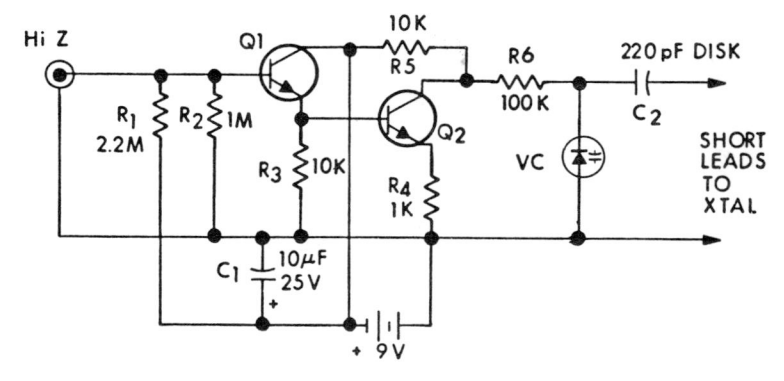

Fig. 2. Varactor frequency-modulator circuit for crystal-controlled AM rigs.

Now, if you don't happen to have a vfo, it becomes a little more difficult to achieve a reasonable level of deviation since it is considerably more difficult to shift a crystal oscillator than it is a vfo. But it can be done without too much work or expense.

The schematic of Fig. 2 is straightforward and should require little explanation. A couple of ideas which might prove useful, however, are as follows:

1. Use the basic crystal frequency which will provide the highest multiplication factor; for example, use a 3, 4, or 6 MHz crystal rather than an 8 MHz fundamental frequency if your oscillator will accept it.

2. A 5-25 pF trimmer across the crystal will assist greatly in centering your transmitter on channel. Remember too, that not every crystal will work in every oscillator and be on frequency. It may be necessary to order a crystal specifically for your rig!

3. Resistor R6 may be varied to suit each individual varactor diode, although the value shown should give satisfactory operation.

Whichever method you decide to use, much pleasure will be derived through the new associations encountered on FM. Remember though, that the only true way to fly is with a full FM system utilizing an FM receiver as well as a transmitter.

Putting the Ninic Pocket Receiver on Channel

Most of us who operate repeaters or remotely controlled base stations wish for things like tiny walkie-talkies and pocket receivers from time to time. A repeater output greatly enhances the capability of any miniature communications equipment. A walkie-talkie through a repeater packs the same punch as the repeater output, and the punch usually comes from the top of a hill or mountain. To be really handy, the transceiver would have to be small enough to be carried anywhere (pocket tuckawayable), and punchy enough to develop a kicky signal from any line-of-sight range. Unfortunately, the cost for such a unit is prohibitive for us poor folk.

But there are units that **do** fill the bill quite nicely. GE makes one. It's transistorized and uses integrated circuitry for added miniaturization. It puts out more than a watt. It's no bigger than two packs of cigarettes. But it costs nearly a thousand dollars.

A pocket receiver is the next best thing. A good miniature FM receiver is small enough for the pocket yet sensitive enough to allow monitoring of the local repeater output from anywhere within the repeater's general range. And to be fully useful as a monitor, it must have a tunable squelch so latent band noise can be eliminated during the no-signal state. GE makes one of these, too. It's called the Message Mate. It's a highly sensitive unit made for high-band paging systems. But it costs more than $200.

So it's pretty understandable that I became pretty excited when I saw a Mann Communications ad for the $74.95 Ninic pocket communications receiver. I **had** to have one! Here's my reasoning:

In my locality, 146.82 MHz is the primary channel. A group of us established a hilltop 146.70 MHz transmitter which is fed all the 146.82 MHz signals heard by a local receiver. If an

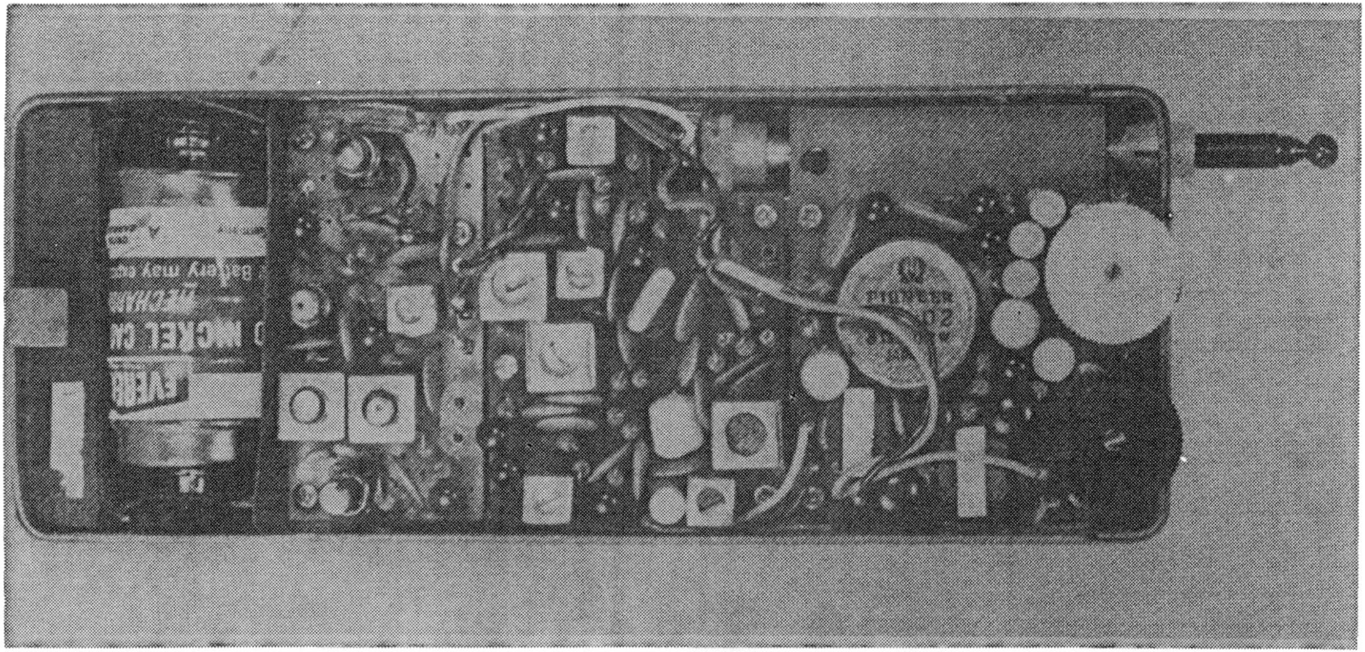

Fig. 1. Photo of open unit shows modular construction and layout.

operator were to crystal up a two-meter mobile unit to transmit on .82 and receive on .70, the mobile would be extremely valuable around town, but totally worthless out of the general repeater area. Similarly, the operator would be out of luck if the .70 transmitter were to malfunction or be forced to shut down for some reason.

An ideal solution would be one whereby FM'ers with single-channel radios could be provided with a means of copying .70 on an auxiliary or secondary receiver. Such a scheme would also prevent a repeater from causing interference.

The perfect "auxiliary" receiver to complement a single-frequency mobile unit by monitoring a repeater output is a sensitive pocket receiver. And the Ninic has proved that it can do the job nicely.

A LOOK AT THE NINIC

The Ninic is a stable, crystal-controlled dual-conversion superheterodyne FM receiver with a sensitivity of about a microvolt for 20 dB of quieting. The unit is inherently noisy, and it takes a signal of greater than 10 microvolts to give complete quieting. The ad said the unit had a sensitivity of 0.3 microvolt. This proved to be true, but there was practically no quieting at that input level.

The Ninic comes equipped with one crystal for the converter and another for the oscillator. The oscillator crystal puts the operating frequency above 150 MHz, but the fact that it is supplied with the receiver greatly simplifies tuneup and checkout; it's always more comfortable to check out a receiver on its original frequency before attempting to set it up on a new channel. The originally supplied crystal allows individual operating characteristics to be observed so you'll know what to expect in the way of performance when the receiver is properly tuned to the new frequency.

The Ninic is comprised of three discrete modular circuit boards, as illustrated in Fig. 1. The smallest of the three is the oscillator module, positioned immediately above the battery.

The low-band (six-meter) receiver is identical with the high-band unit pictured, with the exception of the oscillator module. The oscillator leads are connected with plug-in pins. A low-band unit can be changed to the high-band version (and vice versa) by unplugging the terminal pins, removing the hold-down screws, and inserting the appropriate oscillator module.

Figure 2 is a schematic of the high-band oscillator. While the supply voltage is shown to be 7.5 volts, the receiver works

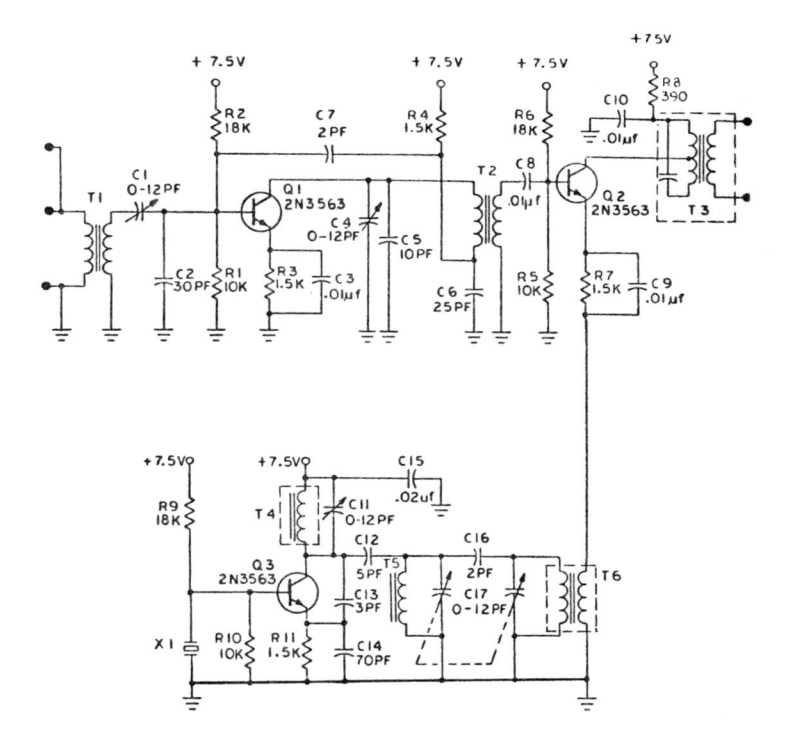

Fig. 2. 150 MHz oscillator module, Ninic pocket receiver.

well with any supply voltage of 6 volts or more. The nickel-cadmium cell in the unit pictured provides 6.25 volts and is more than adequate to allow squelch-breaking on signals better than 0.3 microvolt. The schematic for the low-band oscillator is shown in Fig. 3.

Figure 4 is a schematic of the crystal-controlled i-f module. This is the center circuit board in the photograph of Fig. 1. As noted earlier, this module is the same for high-band or low-band receivers.

TUNEUP

If you're thorough, you will by now have noticed what appears to be a discrepancy between the schematic of the oscillator and the module itself. The schematic shows a number of variable capacitors in the oscillator for tweaking the thing on its new frequency, but even the closest examination of the oscillator won't reveal even a hint of a tunable capacitor. When I saw this apparent disparity, I merely shrugged and muttered some under-the-breath

123

comment about radios made in Japan. And I started the tuneup procedure by adjusting the wax-filled slugs in the coils and transformer cans.

The receiver had all the earmarks of a unit that could be set to any frequency in the two-meter band. I noted that most of the slugs were nearly all the way out. This seemed almost a sure indication that the frequency could be lowered a great deal without modification of the circuit. Then, when two of the slugs bottomed out, I became suspicious. The tuning range of the oscillator was nowhere nearly as broad as it looked to be.

The capacitors in the schematic simply **had** to be on that board somewhere. In desperation, I disconnected the test leads and interconnect pins and pulled the oscillator module out of the receiver. There they were! All the miniature tunable capacitors were protruding from the underside of the oscillator board. This layout makes the tuneup somewhat more difficult from the standpoint of convenience because the oscillator cannot be adjusted while it is mounted in the receiver. The interconnecting leads to the oscillator can be reconnected to the board after it has been removed, however.

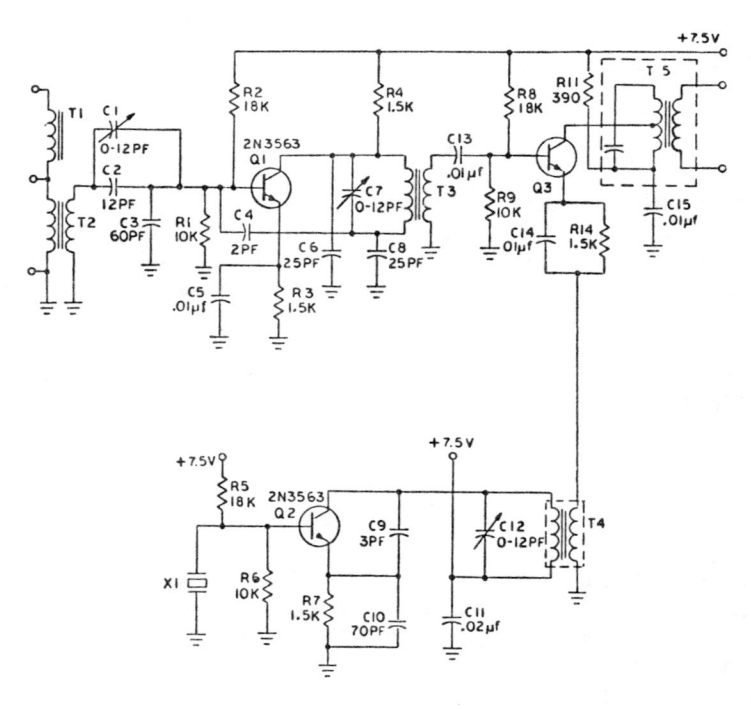

Fig. 3. 50 MHz oscillator module, Ninic pocket receiver.

124

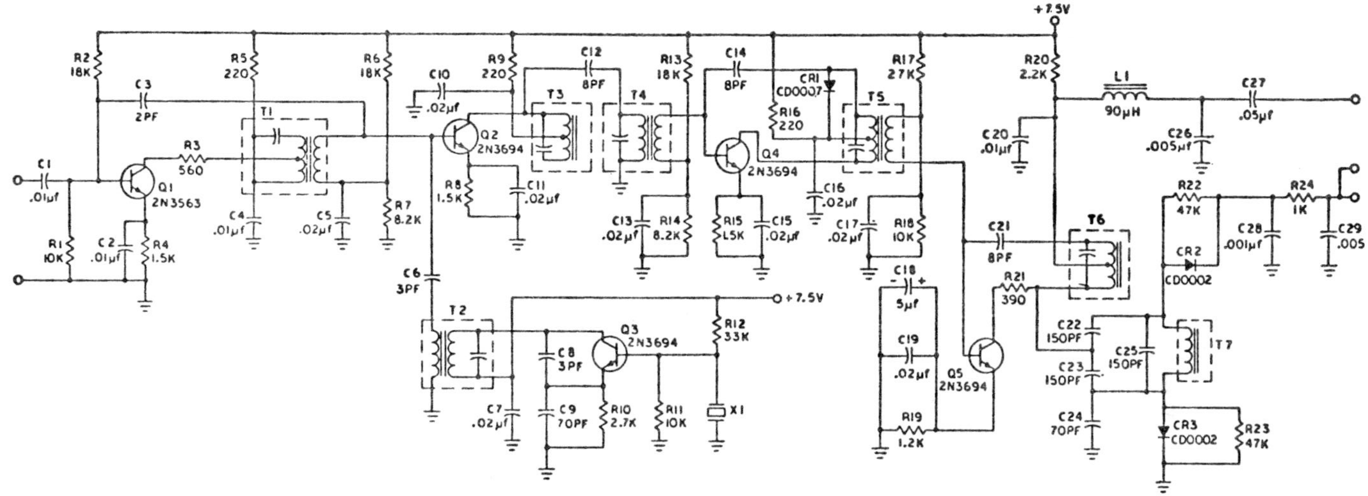

NOTE:

No I-F adjustments are necessary. Adjustments of the slugs and transformers is not recommended. The I-F circuits have been factory-aligned.

Fig. 4. Crystal-controlled i-f module of Ninic pocket receiver.

The board must be placed vertically and situated immediately adjacent to the i-f module so the leads will reach. Fortunately, the oscillator is stable enough so the tuning won't change after the unit is remounted.

Before discussing the tuneup procedure, it might be wise to describe the crystal-ordering particulars. Since Sentry Manufacturing Company advertised regularly in FM—and their crystals really are of the quality we FM'ers have come to expect—I decided to ask them to assign a specific order number to the Ninic oscillator crystal. They **did** do this for the two-meter unit, but because of a lack of information at the time they could not provide correlation data for the six-meter version.

To order a two-meter crystal, give Sentry the two-meter operating frequency desired, specify crystal holder SCM-18, and mention that the receiver is the Ninic.

Here is a complete step-by-step tuneup procedure:

Frequency Adjust (Fig. 5)

    1. After inserting the proper amateur crystal (Sentry SCM-18) remove the three hold-down screws.

    2. Unplug the lead covered with transparent polyvinyl tubing (extreme left in Fig. 5).

    3. Gently lift the lower edge of the board (the end next to the battery) as if opposite edge were hinged.

    NOTE: The oscillator will now be positioned perpendicular to the receiver so adjustments can be made from either side of the board.

    4. Carefully remove the wax from the two transformer cans located below the crystal (see photo).

    5. Using a jeweler's screwdriver, turn both slugs clockwise one complete turn. (On six-meter unit, turn slugs counterclockwise one turn.)

    NOTE: The slugs of all transformers are held in place by tiny one-piece rubber-band segments. If tuning is difficult or if the slug has a tendency to edge back toward the original setting, remove the slugs completely and lift out the rubber-band sections. Then replace the slugs and set as described in step 5.

    6. Leave the polyvinyl-jacketed lead disconnected. This lead attaches to one of the pins on the base of the receiver and allows for the connection of an external antenna.

    7. Connect the output of a signal generator to the antenna lead to produce a saturating signal on the receiver. The squelch should be fully opened.

Fig. 5. Photo showing closeup of oscillator module (note that variable capacitors are on bottom side).

8. Turn on your base station receiver (or any receiver known to be on the same frequency that you're zeroing).

9. Connect a discriminator meter to the base station receiver so it will be in plain view while you're making adjustments on the Ninic.

NOTE: The signal generator output must be high enough in signal strength to be copied on the base station receiver. The base station receiver audio gain should be all the way down.

10. Gradually adjust the signal generator frequency for a zero indication on the base station discriminator meter.

11. Adjust the frequency trimmer capacitor (located in corner on bottom of board below previously adjusted transfromers) until presence of the saturating signal is indicated.

12. If the signal can't be brought in as described above, set capacitor to center of its range and adjust transformer slugs one additional turn. Then repeat step 11.

13. With the receiver saturated as described in step 12, decrease the signal gradually for a popping indication (about 10 dB quieting on the Ninic).

14. Adjust the antenna capacitor for maximum quieting. (This capacitor is located below the open coil at the extreme left.)

15. Reduce signal and readjust capacitor for maximum quieting. Note position of capacitor. If fully open or meshed the antenna coil must be adjusted. Continue adjustment until antenna tuning capacitor provides the best signal near the center of its range.

16. Adjust all other oscillator board capacitors for maximum quieting, but remember to reduce the input signal each time it reaches saturation.

17. Observe all capacitors. If any are fully open or closed, a readjustment of the corresponding coil is necessary.

18. When receiver has been fully tuned, reset all coils with wax. Use an ordinary birthday candle and allow the melted wax to fix the slugs in their new positions.

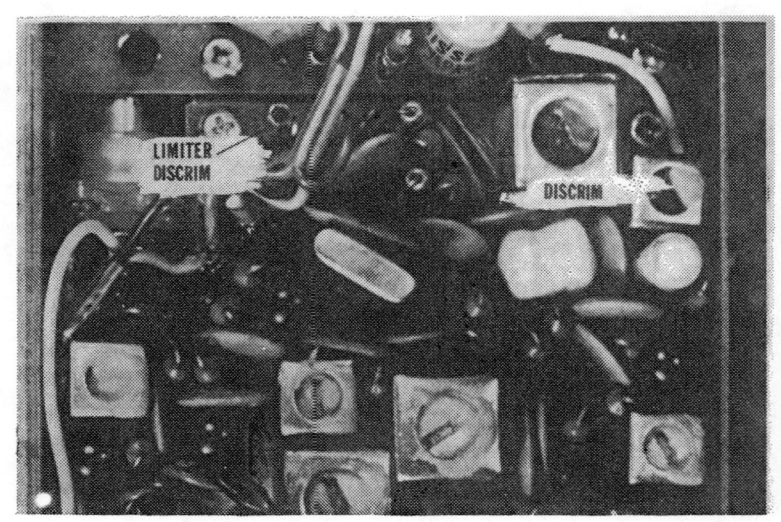

Limiter/Discriminator test point is an all-purpose alignment point, and may be used to monitor limiter current as well as discriminator position. When the input signal is below saturation, use a 2V meter and peak all adjustments (maximum meter deflection). Then, with saturating signal, adjust oscillator for 0 on meter. Make sure discriminator can is adjusted for 0 with no input signal.

Fig. 6. Photo of Ninic i-f module.

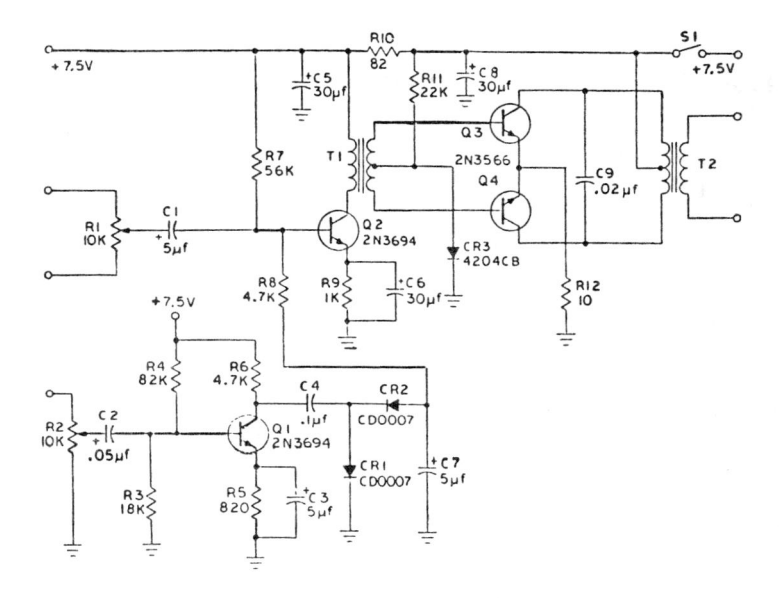

Fig. 7. Audio and squelch module, Ninic pocket receiver.

Final Comments

The discriminator adjustment is in the upper righthand corner of the i-f module. Figure 6 illustrates this and also shows the first-limiter monitoring point. When monitoring for limiter current, set meter to either the 50-microamp scale or a 0-2V dc scale (20,000 ohms per volt).

Figure 7 is a schematic of the squelch and audio circuitry. There are no adjustments to be made on this board; the circuit is included only for reference. (The Ninic does not come with schematics; the originals from which those in this article were obtained were probably the only set in existence.)

Split-Reed Vibrators for Simplified 6- to 12-Volt Conversions

by Bill Harris

Some of the more common and newer pieces of two-way gear lend themselves readily to 12 volt conversion; some by restrapping the cable plug; other types simply by plug-in jumper arrangements. Much of the older equipment, however, was designed as strictly 6 volt, with conversion to 12 generally necessitating the procurement of a new dynamotor as well as a new vibrator transformer and vibrator.

Obtaining a 12V dynamotor is seldom a problem for the average FM operator because many types are available surplus* and some are being retired in order to change to transistor power. Therefore, we will not delve into the dynamotor aspect at this time, but will concentrate on the vibrator supply.

Studying the average 6V mobile unit to be converted, the first item of note is that the existing vibrator transformer is usually in clean condition and does match the chassis space and construction fairly closely. To maintain this ethnic symmetry one would normally want to replace the transformer with one having the same physical characteristics but with the exception of a 12V primary winding instead of a 6. The first thing that comes to mind here is the price of such an item. Since it is an item that doesn't generally wear out, it's a pity to callously throw it away.

The vibrator, however, can generally be considered well on the road to being in need of replacement, since it is an item that does wear out. In addition, some of the original equipment 6V vibrators are getting increasingly scarce and expensive. So if we're going to replace it anyway, why not spend the money on one that will: (1) allow the use of the old transformer, (2) last nearly forever, (3) be easily obtainable for replacement. There is one vibrator that fulfills all these criteria: the Mallory No. 1701 (or Motorola 48B800145). This unit is of split-

---

*Two common types of surplus "dynos" are the Signal Corps DM-35, 12-14V input at 28 to 32 amps; 625V at 225 mA output, good for 829-B and Pr. —807 rigs; and Eicor D-401 (Collins TCS Eqpt), 12-14V input at 18-22 amps, 420V out at 200-240 mA for Pr. —2E26 rigs, etc.

Figure 1

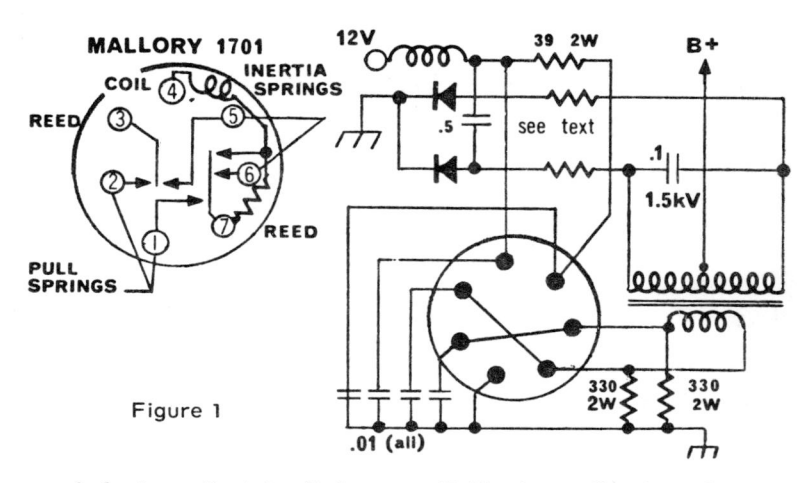

reed design; that is, it is essentially two vibrators in one, which allows us to use the 6V transformer in a 12V circuit. It is of heavy enough design and it is used to power 50W transmitters; hence the longevity. It uses a common 7-pin base, which makes obtaining sockets easier; and it is one of the more common communications replacement vibrators, so it can be found in almost any two-way radio shop or radio parts catalog. The price runs about $7 or $8, which is indeed reasonable considering the above factors.

In case you have heard that split-reed vibrator supplies are noisier than their simple counterparts, don't accept that as gospel. The only causes of excessive noise in a vibrator supply are: poor design or layout, inadequate shielding, lack of bypassing, or a worn-out vibrator. I've never experienced noticeable noise from a split-reed conversion, and I suspect it's because a new vibrator is almost always employed.

If you are ready to convert, here's how:

1. Carefully strip out old vibrator socket from chassis. Save ground cup, buffer capacitors, wire leads and any bus-wire jumpers and "spaghetti."

2. Install new socket (Amphenol No. 77-MIP-7) in hole vacated by previous socket; it may be necessary to file hole out a few thousandths.

3. Ground pin 7 to chassis with a short, heavy bus wire.

4. Connect 12V "hot" lead (or hash choke) to pin 3.

5. Using heavy bus wire and spaghetti, jumper pins 1 to 5, and 2 to 6 of socket.

6. Carefully wire a 0.01 uF disc from chassis ground to each of pins 1, 2, 3, and 4. Use more than one ground if necessary, but keep cap leads short.

7. Connect a 39-ohm, 2-watt resistor from pin 3 (A hot) to pin 4 of socket.

8. Connect one each (total of two) 330-ohm, 1-watt carbon resistor from ground to pin 5 and ground to pin 6 of socket.

9. Connect transformer primary OUTSIDE leads to pins 5 and 6 of socket. Leave centertap lead disconnected and tape off securely! The primary voltage is now applied at the vibrator socket and not at the transformer.

10. Install whatever type of tie-lug terminal strip(s) you can find that will fit in the vicinity of the transformer secondary leads.

11. Build up a bridge or full-wave bank using a minimum of two 400 PIV or higher silicons per leg for each 200 volts expected on the secondary. Use small value carbon resistors (10 to 100 ohms) in series with each secondary lead. Don't omit these protective resistors. Use spaghetti where necessary to insulate diodes and resistors from chassis and other wiring.

12. Recheck all wiring before applying power. Make sure buffer capacitors have been replaced with the proper values.

One caution: Do not operate any split-reed supply without a proper load! Secondary voltage will go sky-high and vibrator will very likely incur damage! When conversion is complete, replace all shield covers, install vibrator, and connect a dc voltmeter of the proper range to the B+ line of the set. Turn on set and observe meter: If voltage is somewhat high, recheck after tubes warm up (10 to 15 seconds). If voltage falls to normal, leave set running and check for excessive vibrator or transformer heat in 5 to 10 minutes. The vibrator should never get so warm it cannot be unplugged with the bare hand; the transformer may run somewhat warmer than the vibrator, though no hotter than it operated on 6 volts.

Improving the Gonset
G-151 FM Communicator
by Bill Harris K9FOV

The Gonset G-151 FM Communicator is a rather late model of commercial two-way gear that may as yet be unfamiliar to amateurs not engaged in land mobile service work, but it does happen to be one of the more common sets in some areas. The G-151 falls under the heading of "licensable" gear, and as such may be used legally in commercial service, although some are appearing on the used equipment market at various prices. This may be due to the fact that it is a rather delicate piece of construction when compared to the more popular makes, and tends to require more frequent maintenance as a result. In addition, the G-151 had a few design shortcomings that have plagued service shops to no end.

This is intended for the enlightenment of the amateur using or contemplating the G-151 FM Communicator, and as a possible help to the two-way service technician who might like to help his clients who have these units achieve more satisfaction from their equipment.

The main problem that I have found with the G-151 has been the receiver; the transmitters seem to have only one major fault which is usually easily corrected with the installation of three new 6360 tubes in the driver and PA stages. The receiver, however, as it comes, generally leaves something to be desired in the way of sensitivity, stability, and squ' lch action.

General Improvement

The first item is to run a complete tube check, replacing any that flunk for any reason. The 6FG5 rf and 6U8A first mixer and multiplier are best bets to watch for shorted or flat condition. These tubes run excessively hot in a stock unit, and as a result don't last long. Correct this by replacing the 47 ohm, ½ watt resistor feeding the rf screen with a 22K or 27K, ½ watt. Incidentally, if you have to replace the 6FG5 and don't have one handy, the old reliable 6AK5 or even a 6CB6 will do the job as well without changes. (Gonset made some fantastic claims about the 6FG5, but in truth the tube was

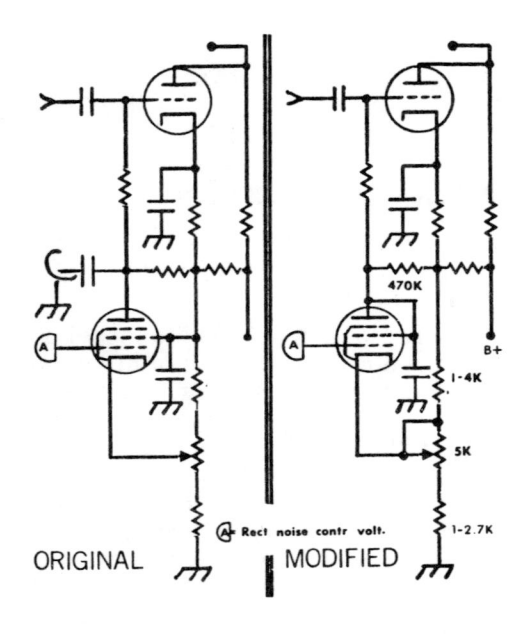

Figure 1

ORIGINAL   MODIFIED

nothing more than a frame-grid version of the 6CB6. At present, all VHF tube types, including the 6CB6, are made as frame-grid types.)

While you have the soldering iron hot, locate the 1K (1 watt) resistor feeding the plate of the 6U8 first mixer from B-plus through the first high i-f can primary. Replace this with a 10 or 15K, 2-watt unit. Changing these resistors should actually increase the sensitivity of the receiver by lowering the noise figure as well as lowering the dissipation of the tubes enough to lower their mortality rate appreciably.

Squelch

Now that we have cooled down the front end, the squelch department is the next stop. In addition to heat caused by excessive current dissipation due to poor voltage distribution, squelch action could best be described as "sloppy" in most units. Now if you are happy with squelch action and the resistors aren't charred too badly, by all means leave it unmodified if you wish. But if you're a tinkerer and would like to try a change, grab the soldering gat and:

Look at the schematic (Fig. 1), which will tell you the squelch—first-audio tube, a 6EB8, is wired as a pentode clamp driving a triode gate. Due to the slow cutoff characteristics of a pentode, this is not an ideal situation. Disconnect the screen wire (pin 8) from the terminal board junction of

R149-R150-R153 and R162 and reconnect it to pin 9, the plate of the 6EB8. This essentially makes a power triode of the pentode section of the tube.

Look also at the plate load resistor of the clamp section of the tube; most models use a 47K. For better control swing, change this to about a 470K (I just wired the new resistor in series with one end of the old one). These two steps should make the squelch act a little more like the high-priced spread, providing you can control it with the front-panel control. If not, change the pot to a 5K (linear taper) of good quality, and you should have the range you need.

As a further improvement, try various values of resistance in series with the ground side of the control (original value in most is a 560-ohm, ½ watt) until you arrive at one that will permit a moderately strong signal to open a "full tight" squelch setting. Some later G-151's had a small screwdriver-adjusted pot behind the front panel for this purpose (shades of DuMont!). In addition, the 4.7K, ½ watt resistor feeding B-plus to the top of the control, located by itself on a strip near the center rear of the chassis, should be replaced if it is charred badly. Experiment here for a value that will improve the "feel" of the squelch control. (I ended up with a 1.5K, 1-watt on the bottom of the pot and 1K 1-watt on the top.)

The apparent quieting sensitivity can be improved markedly by adding a little more deemphasis; it's easiest in the audio output stage. A 0.039 uF Discap from the 12AQ5 grid to ground, or a 0.02 uF at 1 kV or higher across the primary of the output transformer will help the rolloff of unnecessary response to frequencies and noise above 3 kHz.

On aligning the receiver strip, don't touch any of the i-f cans until you have managed to get an on-frequency signal through the receiver and peaked up the various front-end coils. A good idea is the use of a good grid-dipper to put the rf coils "in the ballpark" before attempting final alignment. None of the coils should require padding to reach 147 MHz, but due to the low amount of capacitance range of the piston-type trimmers used, some may be compressed a little in order to lower the tuning range.

After the front end is aligned, recheck the discriminator and make sure the "rubbering coil" near the receiver crystal will move the signal both ways a little; if it won't, check to see whether some enterprising service monkey hasn't soldered a jumper across the coil or clipped the 15 pF silver-mica cap from the terminals of the second i-f crystal socket in order to compensate for a defective original crystal. If either of these defects is noted, correct it.

Carefully go through the low and high i-f strips once or twice and recheck discriminator idling on noise and signal. If you can get them even close, that's good for the G-151. If not, don't worry about it—widebanding will correct that problem to a great extent. After alignment is completed, check for proper squelch action with and without signal. If the alignment or stage gain is not up to par, you may not be able to close the squelch by means of the front panel control. It may close properly but refuse to open even on a full-quieting signal; if one or the other is the case, check the 6EB8 and the 6BN8 tubes for low emission.

Assuming that the receiver appears to operate okay (25 or 30 uA, or about 0.5V first limiter current at the metering socket), the next step is to wideband the dear little thing, that is unless you are planning to use it in conjunction with a system of unmodified TPL's and Motracs. To "unsplit," we'll need the following: one 2-inch piece of insulated hookup wire stripped ¼ inch at both ends; three 68K, ½ watt resistors; one 180K or 220K, ½ watt resistor, and one 150K, ½ watt resistor.

Widebanding

Assuming that you have access to a working Gonset G-151 and you want to put it on say, 146.94, the first step would naturally be the selection and procurement of new crystals. The receiver will take a 45.413333 MHz (146.94 — 10.7 ÷ 3) and the transmitter will go on with a 12.245 MHz crystal. A good point to make here is that some units will not quite make 15 kHz deviation using a 12 MHz crystal, and modifying the circuit will not help this. Using instead a 6,122.50 slab and doubling in the oscillator or modulator (like GE) should allow more than adequate deviation within control range. A good idea would be to check the unit on its original frequency before conversion, providing you have access to shop facilities, in order to ascertain whether it will deviate 15 kHz. If not, have Sentry correlate the 6,122.5 crystal for you instead of the 12.245. Also, at the point of ordering crystals, decide whether you want the oven or not; if so, order small-pin, 85°C oven type crystals. Otherwise get standard commercial-grade 0.093-pin room-temp units. Either type should be comparably stable.

When the crystals make their debut, chonk them in the proper sides of the oven or oven socket as the case may be, then fire the unit up and align the transmitter strip according to the instructions in the service manual. All coils should tune very close to their original settings, although L202 (the 75 MHz tripler plate coil) may have to have the turns compressed a bit closer together. Disregard power output if you get anything at all. Unless the 6360s are almost new, 15 or 20 watts is about par; just make sure you have at least —45 volts drive (use VTVM) at the centertap of the PA grid coil.

Note that there are two 455 kHz i-f transformers coupled back-to-back between the first low i-f amp and the second. Unsolder the 2.2 pF tubular ceramic cap from lug 1 of T104 (the can in the corner); unsolder the lead coming from lug 1 of T105 from the grid (pin 1) of the second i-f tube socket. Connect the 2-inch piece of hookup wire from lug 1 of T104 to pin 1 of the 6BJ6 second i-f socket. This effectively removes T105 from the circuit. A good idea is to tag the top of T105 with a small label saying "do not tune" or something to that effect. This keeps it from adding to the confusion during alignment.

Continuing, connect a 68K resistor across the secondaries of each of the 455 kHz i-f and limiter transformers; put the 150K across the primary of the discriminator can and the 180 or 220K across the secondary winding. All these resistors are installed at the transformer terminals externally; there is no need to remove the cans from the chassis. This completes the electrical "widebanding" of the G-151 Communicator.

Upon completing the above procedure, careful realignment of the complete receiver is a good idea. Use your favorite method, and repeat the process until you are satisfied with the sound of the unit. Again on the squelch, if operation is sluggish or control will not close squelch, jumper across the 2.2K resistor in the cathode circuit of the noise amplifier (6BN8) tube. This will increase the noise amp again.

Even after all modifications, the G-151 must be used carefully for best results. Avoid overtightening the squelch, and make allowance (or corrections) for overdeviating signals. The unit will accept 15, and that's about it! Anything more and the distortion is really severe.

The Gonset G-151 FM Communicator is a fine "little tiny radio," and will prove adequate for the job it was intended to perform. As long as you don't insist on mounting one on a Motorcycle or a Le Tourneau, it should stand right up alongside the "big boys" in performance and convenience. Let's dig 'em out and get 'em on!

Two-Wire Remote with Zener-Stabilized Squelch

by P. J. Ferrell

One way to keep a radio system simple enough to be controlled from a single pair of wires and reliable enough to be operated by remote control is to eliminate the adjustable squelch of the receiver to be used. The most common problem with fixed-setting squelch systems is "drift." During hot weather the squelch on the remote receiver may open or otherwise operate erratically. In the cold of night, the squelch may lock down tight enough to keep the receiver from hearing weaker stations. This squelch drift can be eliminated through the incorporation of a simple zener-diode stabilization scheme. Zener stabilization eliminates squelch drift and allows instantaneous squelch recovery when the remote station returns to the receive mode.

As shown in the typical squelch circuit of Fig. 1, the zener (about 30V breakdown) replaces the R/C network at the cathode of the squelch amplifier.

Figure 2 shows a method by which a single wire pair can be used to control a remote transmitter/receiver combination. The dc control voltage should be held down to no more than is required to key the current-operated relay (K1). This relay is a DPDT type used to key the transmitter and switch audio between transmit and receive. The circuit within the dashed-line area shows a network which can be used to replace the jumper at AB in case on/off control of ac power is

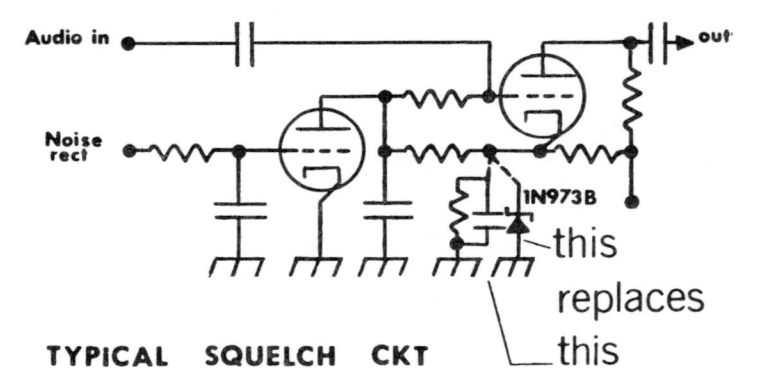

Fig. 1. Zener diode replaces /C net to stabilize squelch.

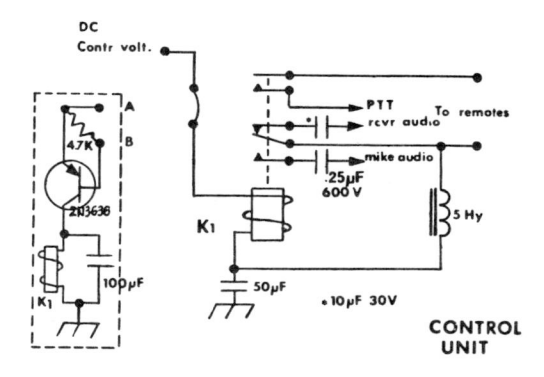

Fig. 2. Using a single pair of wires for controlling remote functions.

required or if a sufficiently sensitive relay is unobtainable. The relay coil specifications are the same, but the enclosed circuit enhances the relay sensitivity.

The remote unit itself is shown in Fig. 3. Several of these remotes can be paralleled, but keep the total wire capacity below about 0.05 uF. This plus the high audio voltage specifically precludes use of conventional telephone lines for control. Speaking of phone lines, keep the remote wires away from your phone wiring or you'll have real live ham talk on your phone. Keep the level of the line down to about 0 dB if possible.

Nearby rf (broadcast, low-band hams, etc.) can modulate your transmitter, too, but an rf choke and a 0.001 uF capacitor at the control unit eliminates the problem super-quick.

Channel changing? How about automatic receiver selection such as shown in Fig. 4. This modification uses FET's as dc amplifiers freeing a triode as the other audio amplifier. Either receiver can be given priority with zener diode Z2. Or a "first-on-wins" capability (not shown) can be obtained by cross connecting another zener opposite Z2.

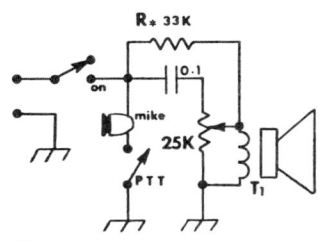

Fig. 3. Circuitry for control end of two-wire line.

T_1 = plate-to-vc

R* omit if not switching ac pwr

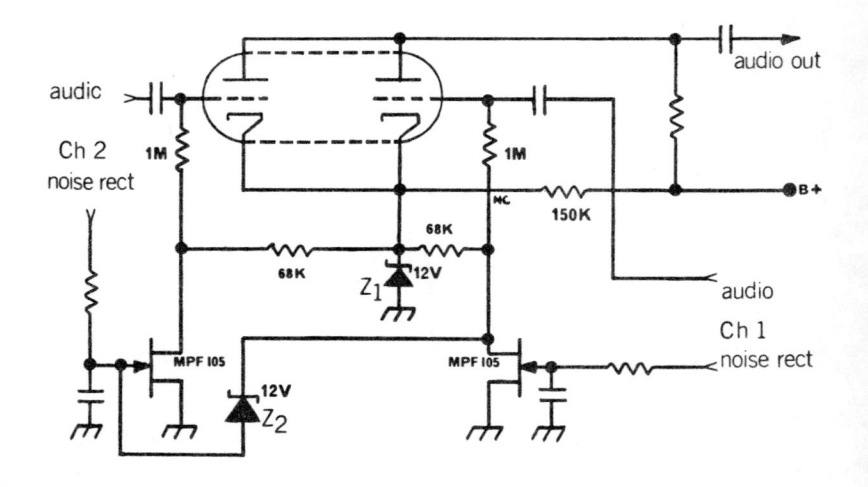

Fig. 4. "Priority select" circuit for remote applications.

Switches in series with these diodes allows for a certain flexibility of response.

Lastly—and having absolutely no connection with the remote capability—we present a carrier-operated lamp circuit (Fig. 5). This is a must for the flashing-lights set; it has the nice feature that lamp brilliance is proportional to quieting for signals in the vicinity of receiver threshold.

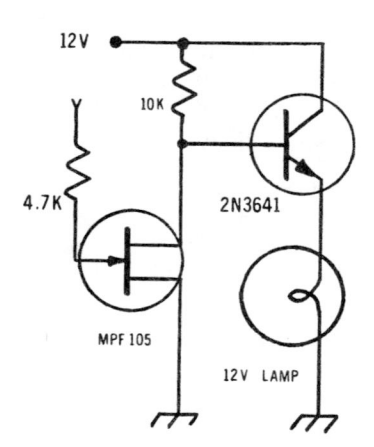

Fig. 5. Lamp driver circuit.

140

Converting the 450 MHz
Prog Line Telephone Mobile

Because of the recent FCC land mobile rule changes for commercial users, there should be an abundance of 450 MHz Progress Line mobile units removed from service. A great many will be special mobile units set up for telephone service. Since these are equipped with Secode selectors, they are not usable "as is" on amateur systems.

But with a little time and a few simple conversions, the Progress Line telephone mobile can be made into a fine amateur unit.

The first thing to do will be to unplug and remove the Secode selector. This is held to the frame by four bolts. Next unplug and remove the transmitter strip. This will give access to the front plug and the T-power side plate.

Very carefully, cut the wires to the cord from which you unplugged the selector (from the plug on the front plate). Three of these wires lead out of the T-power unit. Save as much of these three leads as you can so you can attach them up to the front plug. The shield wire goes to pin 19 on the plug, the red wire to pin 20, and the black wire to pin 21. There is a fourth orange wire feeding from the T-power to the selector cable. This is a B-plus wire. Cut and tape it.

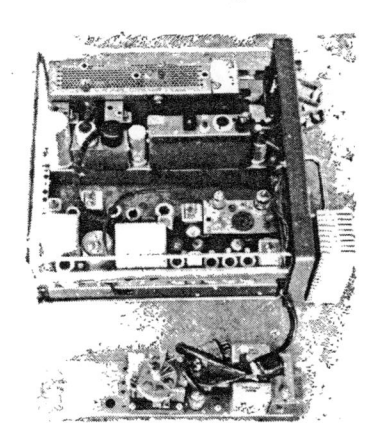

The T-power GE 450 Progress Line with Secode strip.

Next, remove the side cover from the T-power strip. This is done by removing the three cross-recessed retaining screws from the top. The cover will then slide up and out.

NOTE: On top of the unit there is a control pot with a large washer. Remove this pot and all the connecting parts that go to the terminal board.

Disconnect the audio cables from the pot and wire cables together. (The wires are color-coded, so this should be no problem.)

If you wish, you can install a standard GE chassis mike connector in the hole from which you removed the pot and wire the cable to this for easy operation directly from your unit. The unit is now ready for tuneup and operation.

Sometimes these units will experience a T-power whine. If this should occur, turn the unit over. Note under the transmitter there is a ceramic trimmer in the oscillator circuit. Pad the trimmer with a 10 to 15 pF NPO capacitor until you can get a reading on the multiplier 1 test jack of less than 0.5 volt. This will, in many cases, eliminate most of this problem.

Two-Freq. and Simultaneous Monitoring with the GE 4ER6

by Bill Harris K9FOV

Probably the most common set of GE two-piece gear on the amateur FM scene consists of the Pre-Progress transmitter in the ludicrous green case, series 4ET5 or 4ET6 (30 watt and 50 watt, respectively) and the 4ER6 (series A through F) receiver. Many of these units were used in systems that use semiduplex or "three-way" frequency setups. That is, the mobile transmitters are capable of two-frequency operation for either calling the base station or other cars in the system at will, while all the mobile receivers monitor the base frequency only. Most police and utility company radio systems are set up in this fashion, so radio sets that come out of such systems usually have a two-channel transmitter and a single-channel receiver. This may be fine for localized repeater use or simplex use (simply ignore the second crystal socket and leave the switch on the control head permanently in the correct position), but many hams will want to be able to duplex between two adjacent channels, and for this the receiver will have to be selectable between two crystal-controlled channels along with the transmitter.

    The first thing that comes to mind is that the transmitter crystals are switched by means of a relay. Why, therefore, can't the receiver crystals be switched in the same manner? Well, a few of the ER6 receivers did indeed have a relay selecting between a pair of crystals. However, the relay was a special low-capacity type and was mounted in such a manner that lead inductance and capacitance were kept to a bare minimum. In addition, the author has not seen any receivers of this type above 43 MHz using the relay. This method of switching was tried in an older 4RMV35 receiver using 47 MHz crystals, and, true to form, it was impossible to get both crystals to exhibit uniform activity, even by trying several types of relays, different hookups, and swapping crystals between sockets. This helped foster the notion that relay switching of receiver crystals in the 6-meter receiver was impractical. With that in mind, I cleaned up the mess in the octal-tube receiver and focused my rheumy eyes on a 4ER6B2 with intentions of dual-frequing it. I came up with the following procedure.

Remove the oscillator compartment cover and peer in. You should have a ready-made hold for another crystal socket right beside the existing one. Fill it with another ceramic socket. There are two small holes about ¼ inch apart directly behind the oscillator meter jack. Imagine another hole, in line with these two but an additional ¼ inch nearer the high i-f amplifier tube, and punch at that point for a 7-pin miniature socket. Drill the saddle holes so that pin 1 points toward the corner of the chassis. When mounting the socket, put a ground lug under the mounting nut nearest pins 3 and 4.

Remove the oscillator grid metering jack. Set aside both it and the 22K meter shunt resistor connected between it and ground. Connect the loose end of the 220K grid resistor to the grounded side of the original crystal socket. Find a small slug-tuned coil form that will fit in the hole vacated by the meter jack. (The store-bought ceramic type such as the Cambridge Thermionic CSTC will fit directly, or if you rob an old ¼-inch fiber form out of a junked TV i-f strip like I did, the hole will have to be filed out a few caliber.) Wind on five or six turns of No. 20 or 22 enameled wire and put an 18 or 22 pF NPO ceramic or silver mica across it. Stick it in the hole and wire one end to pin 1 of the new socket. Run the other end of the winding directly over to the B+ end of the existing oscillator coil. (Even though this lead is quite long, there seemed no need to bypass it at the coil.)

Connect a 1500 pF 10 percent disc capacitor between pin 7 of the new socket and pin 3 of the 12AT7 (first mixer). (This is also a long reach; use spaghetti on the leads of the cap.)

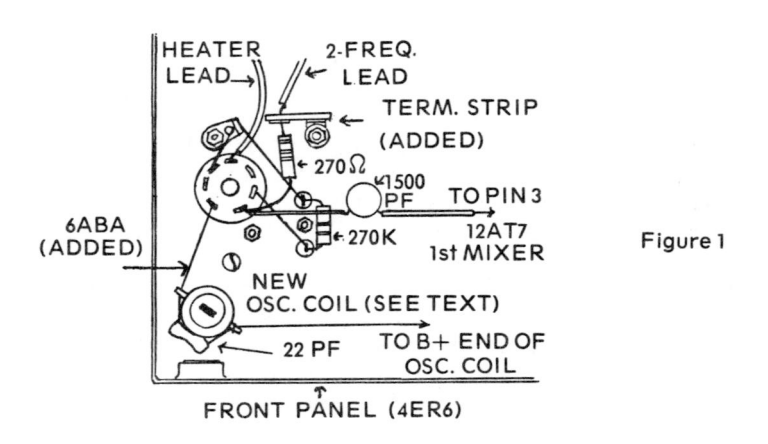

Figure 1

Connect one end of a 270-ohm resistor (half-watt) to pin 7 of the new socket and run the other lead of this resistor to the ground lug. Feed the leads of a 270K ½-watt resistor through the lugs of the new crystal socket and connect one to pin 6 and the other to the ground lug. This completes the rf wiring; all that remains is to connect the heater and cathode circuits.

This is the point at which to decide how we are going to switch between channels. On most of the Pre-Prog units there were just enough wires in the control cables to go around. However, some have a high-impedance volume control on the chassis instead of at the head, and this leaves an additional lead unused in the cable. In these units, also, a red wire was reserved for two-frequency operation, and this should give you a pair of unused wires. In this case, you can wire the oscillator cathodes to ground through the switch, Motorola style, by removing the cathode resistor of each oscillator, bypassing it with a 0.01 uF disk, and sending it up to the head to be grounded by the switch. This provides a "1 - 2 - BOTH" operation with the use of the proper switch.

Let's say you have the older unit that has a high-impedance control at the head. In that case, all the wires are used up and no two-frequency lead is provided. Chin up—there's still hope! You can disconnect the capacitor coupling the output of the discriminator to the control socket connection that goes to the top of the volume pot, and reconnect it to the socket lug that comes back from the rotor of the pot. This effectively converts the volume control from a series to a shunt type and you should not notice any difference in its operation. It works fine for Comco, I might add. They wire theirs that way.

In the head, clip the wire off the top of the volume pot and wire it to the two-frequency switch. Back in the receiver, run a lead from the reclaimed lug on the control receptacle to the oscillator compartment.

Now, how are we going to electrically select the oscillator tube we wish to use? If we have come back with two wires, as in the latter receiver, we either connect the cathodes to the leads, instead of to the chassis, and select ground with the switch in the head (changed to a DPDT), or even select which heater to turn on, if the 11-second delay would not be objectionable. This could be done with either the hot or cold side of the heaters. But if we have the set with only one wire coming back, we will have to do something a little different. Either use a small relay to select cathodes or filaments, or allow the original oscillator to run all the time and turn the new oscillator on and off by means of operating the cathode or filament through the single lead. This has the disadvantage of not being able to mute the calling channel while receiving on the secondary one, but in areas of light activity this might not be too objectionable. In addition, I have found that the GE receiver exhibits an almost imperceptible loss in sensitivity while simultaneously monitoring two channels (I used a Measurements 560 generator for these determinations), so loss of sensitivity should not be of concern in this case.

As for hooking up the heaters, we will assume the reader attempting this modification has done a conversion or two and knows how to handle 6-12V strings. The 6AB4 tube draws 150 mA at 6.3V and can be seriesed with either a 39-ohm 2W resistor or a No. 47 bulb for 12, or tied in parallel with the other tubes for 6.

The missing meter jack may cause feelings of insecurity in case of a breakdown in the field, but is not necessary once the oscillators are initially set up on frequency. Use a VTVM on the 50-volt negative range and meter the respective crystal while tuning the coil for a peak. Then back the slug off to 90 percent of that reading on the slow side. If the new coil does not tune to the crystal frequency, check all the connections at the 6AB4 socket and take a dipper to the coil to see where it tunes in the circuit with the tube in its socket but the power off. Modify the coil if necessary to achieve oscillation.

FM on 10

Joel Eschmann K9MLD

Yes, believe it or not, there is FM activity on 10 meters. Years ago numerous local hams, including myself, acquired some GE gear (4ER6A5 and 4E6A5, transmitter and receiver, respectively).

Now a decision had to be made. Were we going to put them on 6 meters or 10 meters? It was decided 10 meters would do well because: (1) we could get good local communications; (2) we could get DX operating on 29.6 MHz; (3) TVI would be at a minimum as compared to 6 meters. The 4ET6-4ER6, though not narrowband units, would demonstrate good adjacent-frequency rejection from stations operating close to 29.6 when skip is in.

The General Electric 4ET transmitters come in two power ranges. The 4ET5 is a 30-watt unit and 4ET6 produces 50 watts output. They come in three frequency ranges: 4ET6A4, 25-30 MHz; 4ET6A5, 30-40 MHz; 4ET6A6, 40-50 MHz. Depending on the unit you have or purchase, the frequency of operation and power supply (mobile or fixed) will determine the extent to which the transmitter is to be converted. To clarify all the different changes would take a 100-page manual. Included here are typical examples of some conversion charts and voltage levels; you and your junkbox can do the rest.

TRANSMITTER

The transmitter I am using is the 4ET6A5 which will tune 30-40 MHz. (See Fig. 1.) It was really not necessary to make any changes operating so close to 30 MHz (29.6), but I decided

The 4ET6A5 wideband GE transmitter is shown (uncased) beside the 4ER6 receiver.

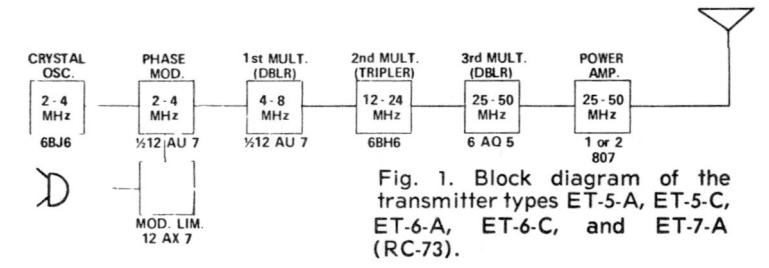

```
CRYSTAL      PHASE      1st MULT.   2nd MULT.   3rd MULT.   POWER
 OSC.         MOD.       (DBLR)     (TRIPLER)    (DBLR)      AMP.
 2-4          2-4         4-8        12-24       25-50       25-50
 MHz          MHz         MHz         MHz         MHz         MHz
 6BJ6       ½12 AU 7   ½12 AU 7      6BH6        6 AO 5     1 or 2
                                                              807
             MOD. LIM.
              12 AX 7
```

Fig. 1. Block diagram of the transmitter types ET-5-A, ET-5-C, ET-6-A, ET-6-C, and ET-7-A (RC-73).

to modify the unit slightly to allow the tuned circuits in the transmitter to function at maximum Q and gain. This means good rejection of any unwanted harmonics getting to the tank circuits.

Table 1 and the schematic of Fig. 2 illustrate the capacitor changes to allow retuning the transmitter to a lower frequency in my case. Just paralleling capacitors and minor tuning was all that was necessary to change frequency of operation. With a multiplier of 12 from oscillator to output, a crystal frequency of 2466.666 kHz (HC-17 / U) was required for 29.6 MHz operation. The filaments require 3.5 amps at 6V and another 1.5 amps to activate the antenna relay. High voltage required was from 500 to 750 volts (250 mA).

Before applying power to the transmitter, use a grid dip meter to tune the first, second, and third multipliers to maximum dip on the multiplier frequencies. If you don't have a grid dip meter, you can tune up with a meter inserted in the test jacks. Although the latter is done anyway, the grid-dip tuneup could be eliminated, but that is not recommended (as described in the tuneup procedure). Now let's tune up the transmitter.

CAUTION: Make certain the antenna or dummy load is connected before tuning.

Table 1. Component Equivalence Chart.

	25-30 MHz	30-40 MHz	40-50 MHz
Model No.	4ET5A 4	4ET5A 5	4ET5A 6
Item Numbers and Values	C110 (120) C113 (120) C117 (47) C118 (47) C122 (39) C153 (91) C156 (180) L112 L115 Y101 Y102	C142 (68) C143 (75) C144 (27) C145 (22) C150 (18) C154 (47) C166 (150) L112 None Y103 Y104	C146 (47) C147 (47) C148 (15) C149 (10) C152 (3) C155 (22) C133 (100) L113 None Y103 Y104

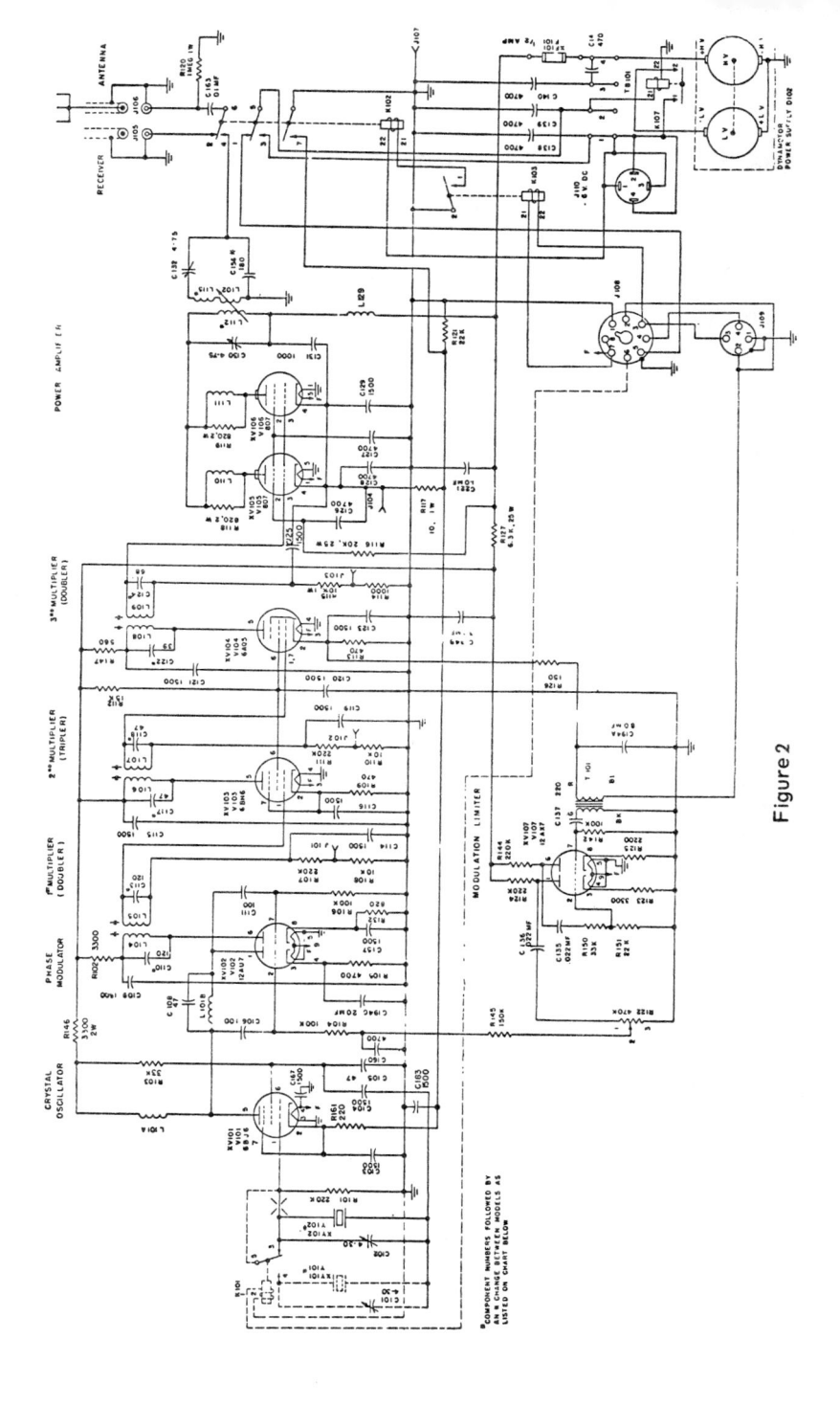

Figure 2

The locking nut on the power amplifier plate tuning capacitor shaft is in the plate voltage circuit. This nut must not be tightened when the dynamotor is running.

The power amplifier housing is intended to protect you from accidental contact with the plate voltage; do not tune or operate the transmitter with the shield removed.

TUNEUP

1. Turn the transmitter filaments on and allow a 30-second warmup period.

2. Disconnect the power amplifier plate connectors. With these connectors off, the transmitter should not be operated more than 15 seconds in every half minute.

3. Connect a 1 mA dc meter between the second multiplier grid (jack J101, the green jack adjacent to the 6BH6) and the chassis (black jack, J107, on the front panel).

4. Key the transmitter and tune the first multiplier double-tuned plate circuit for maximum grid current in the following stage. Best results are obtained by tuning the grid coil first. The maximum grid current will be between 0.1 and 0.3 mA.

5. Insert the 1.0 mA dc meter in third multiplier grid jack J102 (green jack adjacent to the 6AQ5) and tune the second multiplier plate coils as in step 4 above. The maximum grid current will be between 0.2 and 0.6 mA.

6. Connect the 15 mA dc meter between power amplifier grid jack J103 (green jack on front panel) and the chassis (black jack on front panel) and tune the third multiplier coils as indicated in step 4. The maximum power amplifier grid current will be between 5 and 15 mA.

7. Connect a 300 mA dc meter (with less than 0.5 ohm total resistance) between power amplifier cathode jack J104 (red jack on front panel) and the chassis.

8. Reconnect the power amplifier plate connector or connectors.

9. Detune the antenna circuit to minimum capacity.

10. Adjust antenna coupling control L102 to minimum coupling (extreme counterclockwise position).

11. Tune power amplifier plate tank C130 for minimum cathode current.

12. Retune the third multiplier plate and power amplifier grid circuits for maximum power amplifier grid current. This maximum will be lower than that measured with the PA plate connections removed. At the conclusion of each of the tuning operations, a check should be made with the grid dip meter or absorption wavemeter to determine whether the multiplier circuits are tuned to the proper crystal harmonic.

13. Advance antenna coupling control L102 to approximately 20 percent coupling.
14. Adjust antenna tuning control C132 for maximum PA cathode current.
15. Advance antenna coupling control L102 until rated full load is obtained. Rated full load for transmitter ET5 is 100 mA PA cathode current. For transmitter ET6, the PA cathode current is 200 mA for full load.
16. Recheck antenna tuning capacitor C132 and PA plate capacitor C130 for accurate tuning.

When the PA plate caps are connected and it is desired to retune or touch up the transmitter alignment under full load, the observed multiplier and PA grid currents will decrease and may fall below the limits given per instructions above.

Under these conditions the lower limit of the readings should be as follows:

Third multiplier grid (J101)	60 uA
Third multiplier (J102)	150 uA
PA grid	3 mA for 30-watt transmitter
	5 mA for 50-watt transmitter

My description and drawings of the transmitter are of early revisions. Due to a design change, only minor layout changes and the addition of a low-pass filter will be evident in later models. It is important to bypass this filter for 50 MHz operation.

RECEIVER

The General Electric 4ER receiver (Fig. 3) was designed in three frequency ranges: 4ER6A4, 25-30 MHz; 4ER6A5, 30-40 MHz; and 4ER6A6, 40-50 MHz. The receiver requires tuning only the front end to facilitate change of frequency unless the unit requires servicing. For following stages, only tuning is required, unless you are making drastic frequency change (such as 10 MHz or so). Table 2 and the schematic of Fig. 4 show component changes for large frequency conversions. A 23.6 MHz HC-6 / U crystal is required for receiver operation. The supply should have a capability of 6 amperes at 6.3V and 180 volts at 80 mA.

When desired, the frequency of the receiver can be changed by changing the high frequency oscillator crystal and retuning the oscillator and rf stages. The procedure for making these adjustments is given below. If the receiver needs to be completely retuned, use the procedure outlined.

Equipment Needed:
1. A nonmetallic screwdriver.
2. Two meters (0-50 and 0-250 microamperes).

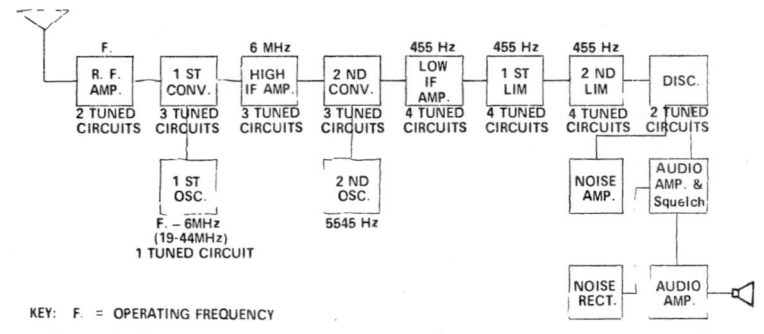

```
     F.         6 MHz             455 Hz  455 Hz  455 Hz
  ┌──────┬──────┬──────┬──────┬──────┬──────┬──────┬──────┐
  │ R.F. │ 1 ST │ HIGH │ 2 ND │ LOW  │ 1 ST │ 2 ND │      │
  │ AMP. │ CONV.│  IF  │ CONV.│  IF  │ LIM  │ LIM  │ DISC.│
  │      │      │ AMP. │      │ AMP. │      │      │      │
  │2 TUNED│3 TUNED│3 TUNED│3 TUNED│4 TUNED│4 TUNED│4 TUNED│2 TUNED│
  │CIRCUITS│CIRCUITS│CIRCUITS│CIRCUITS│CIRCUITS│CIRCUITS│CIRCUITS│CIRCUITS│
```

Fig. 3. Block diagram of the mobile and station receivers types ER6A, ER6B, ER 6C, ER 6D, ER 7A, and ER.7B (RC-181).

3. A crystal of the proper frequency for the high frequency oscillator.

4. Signal generator (25-50 MHz ranges).

To change frequency or align the rf and antenna stages:

1. Turn the receiver on and allow it to warm up for two or three minutes.

2. Insert the new crystal in the high-frequency oscillator crystal socket.

3. Insert the 0-50 microammeter in oscillator grid jack J302.

4. Tune oscillator plate tank coil L301, L302, or L303 for maximum oscillator grid current.

Table 2

Model No.	4ER6A4 & 6C4		4ER6A5 & 6C5		4ER6A6 & 6C6	
Component Symbol Numbers	C361	1.0	C375	1.0	C376	1.0
	C362	1.2	C377	1.0	C378	1.0
	C364	39	C379	18	C380	27
	C365	43	C381	18	C382	22
	C366	36	C383	7	C384	39
	C367	33	C385	22	C386	51
	C368	43	C387	15	C388	47
	C372	68	C373	33	C374	18
	L301		L302		L303	
	T301		T321		T331	
	T302		T322		T332	
	Y301		Y302		Y303	
	R393	22K	R384	33K		
	R394	33K	R385	33K		
	R395	22K	R392	33K		
	C307		C307		C434	
	C308	4.7-45	C308	4.7-45	C435	4.7-53
	C309		C309		C436	

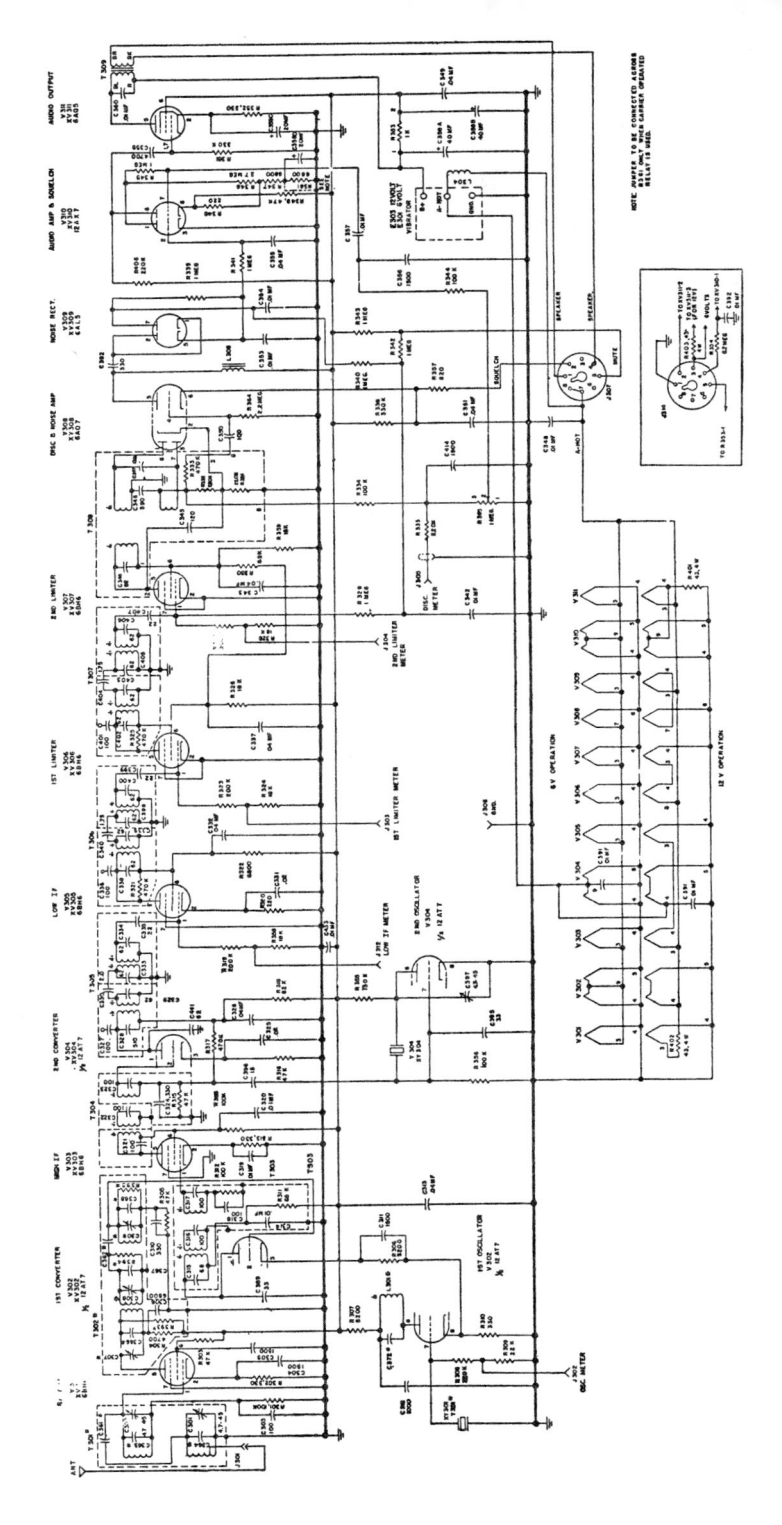

Figure 4

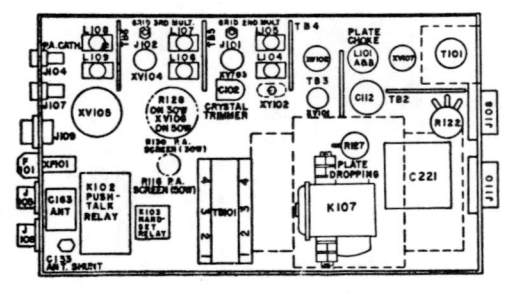

A Outline Diagram of the Transmitter Models 4ET5A4, 5, 6; 4ET6A4, 5, 6; 4ET5B4, 5, 6; 4ET6B4, 5, 6, bottom view (RC-223)

Figure 5

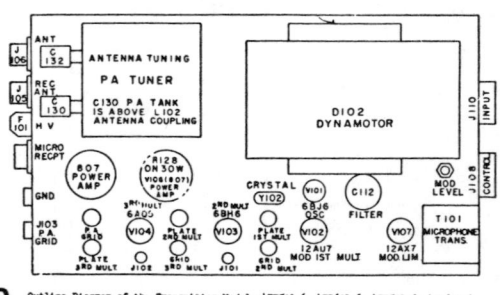

B Outline Diagram of the Transmitter Models 4ET5A1-6, 4ET6A1-6, 4ET5B1-6, 4ET6B1-6, top view (RC-77)

5. Note the reading and turn the iron core counterclockwise until meter reads 50 percent of the maximum grid current (approximately one-half turn).
6. Insert a 0-250 uA meter in first-limiter grid jack J303.
7. Apply an unmodulated signal to pin 1 XV301, the first rf grid, through a 0.01 uF capacitor.
8. Peak transformer T302, T322, or T332, the first rf plate transformer, for maximum first-limiter grid current. Start at the bottom trimmer and work toward the top.
9. With the receiver connected to the proper antenna, transmit on the operating frequency a weak, unmodulated signal from the signal generator. Keep the signal level low enough so that the limiters will not saturate.
10. Peak transformer T301, T321, or T331, the antenna transformer for maximum first-limiter grid current.

I have noted many FM surplus centers around the country are selling these GE rigs with control heads and cables for around $50. I would recommend that if you are going to purchase any units, specify the model which would suit your frequency of operation and, of course, a 6 or 12V unit for

154

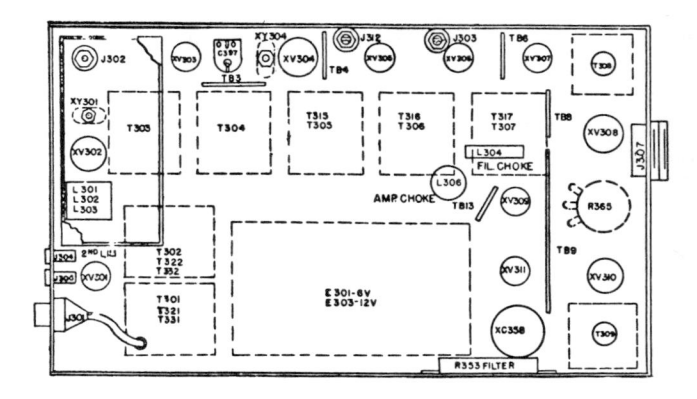

Fig. 6A. Outline diagram of the receiver types E-R-6-A, B, C, D (G-4, 5, 6) (RC-179).

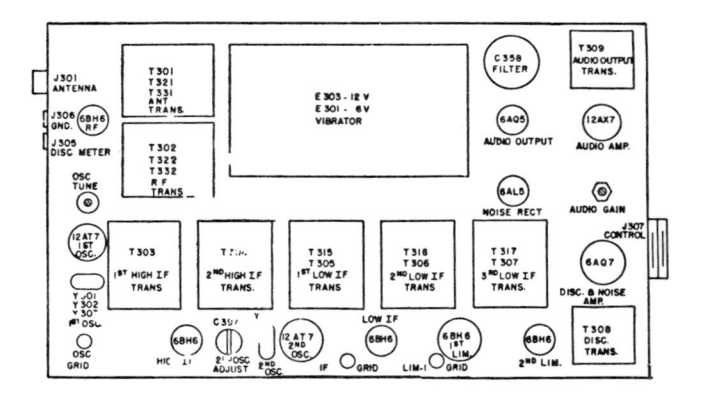

Fig. 6B. Outline diagram of the receiver types. ER-6-A, B, C, D, (G-4, 5, 6) (RC-180).

mobile or fixed. Figures 5 and 6 show outline diagrams for various transmitter and receiver types.

I have in operation two units, one mobile and one fixed. The mobile transmitter unit derives its high voltage from a dynamotor supply. To increase the transmitter efficiency, I built a transistorized converter to give the necessary high voltage upon replacing the dynamotor. Keep an eye on the surplus market for converter transformers. In the interest of completeness, Fig. 7 shows the mobile interconnection data. For the fixed installation, television transformers do very nicely to supply the high voltage required and with the dynamotor and dynamotor relay removed from the chassis, there is ample room for these components. Fixed installations require no control head. Removing the power connectors and adding small aluminum plates to the end of the chassis, you can mount the microphone socket and power switch for the transmitter, squelch, and volume in the receiver. In the receiver chassis, there is sufficient room for a speaker if a power transformer is placed close to 6AQ5. Cut a hole in the side of the cover and mount the speaker grill.

Well, with the conversion complete, plug in an old CB ground plane (shortened, of course). We'll be looking for you on 29.6 MHz!

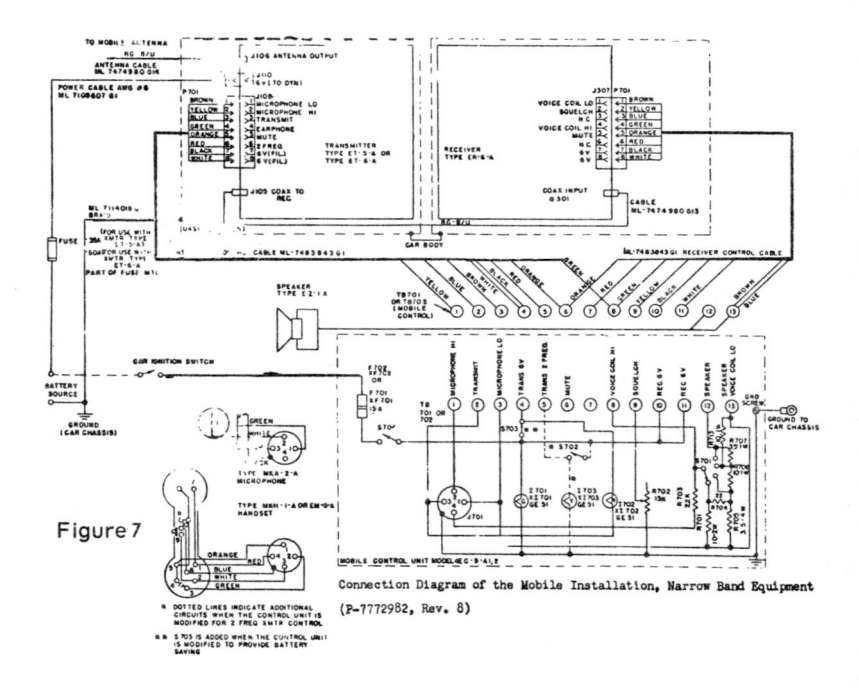

Figure 7

Connection Diagram of the Mobile Installation, Narrow Band Equipment
(P-7772982, Rev. 8)

Complete Narrowbanding of the GE Pre-Prog
by Jim Lev K6DGX

Serious FM operators are constantly refining and improving their equipment; witness the incredible multihop repeaters of the East Coast and the many fine remote bases scattered throughout the west. Consider the heavy demand for articles on Touchtone control, T-power supplies, transistorized accessories, soup-up info, etc. Never before in the history of FM amateur radio have so many been doing so much. Who knows, if present trends continue, it just may be that Motorola and General Electric might come knocking on our doors (hat in hand) for new ideas and techniques! Some might question my definition of the word "improve"; to them I offer the following: "The amateur is constantly in the forefront of technical progress. His incessant curiosity, his eagerness to try anything new, are two reasons." The foregoing is a direct quote from the Radio Amateur's Handbook.

Narrowband operation on 450 MHz is something new; only recently the big names in two-way radio introduced new narrowband 450 MHz equipment, and across the country dozens of service shops worked hard to meet FCC narrowband deadlines. This should be of special interest to those in the profession who desire to retain the venerable GE Pre-Progress MC306 receiver in repeater service. Commercials are narrowbanding out of sheer necessity and not for many of the reasons that we would undertake such a step. To them it is a business; to us it is a hobby that we love.

There are advantages and disadvantages to be considered when thinking of a switch to narrowband; I will cover some of them here. The GE MASTR Line 450 MHz receiver is a fine piece of engineering that is available either wideband or narrowband; interestingly enough (and despite identical front ends), GE claims a 20 dB quieting sensitivity of 0.65 uV for the wideband version and 0.40 uV for the narrowband one; it is doubtful that this 4 dB difference can be attributed to narrowbanding, but I have noted an increase of approximately 2 dB in quieting sensitivity on many of the 100 or so Pre-Progress Line receivers that I have narrowed. So, as a result of narrowbanding, you may gain a few dB; this is nothing to scoff at as decibels are increasingly hard to come by as your repeater system approaches the "ultimate."

Another point: What about lessened interference from adjacent-channel operations? If you are now being bothered by your neighbor, tightening up your receivers can help a lot; also, if you are plagued with intermod, narrowbanding could be your answer. Furthermore, if you now live in a congested 450 MHz area (like Los Angeles) and use two channels side by side, how would you like a "free" third channel right in the middle? In the Los Angeles area and probably elsewhere in the country, 450 MHz activity has increased to a point where virtually every 50 kHz channel between 440 and 450 MHz is in use. Are you naive enough to believe that your rival across town or your buddy down the street is going to be dissuaded from building his own remote simply because all the channels are in use? Hah! Would you have been? You know where he is going to put it, don't you? Probably right between you and what's his name! If your receivers are broad you can say goodbye to the good old days of a clear channel. Commercials are narrowbanding because of congestions, and we amateurs may eventually be forced to do so for the same reasons.

Besides all this, aren't you curious to see if you can do it? If for no other reason, wouldn't you be proud to number your remote among the pioneers across the country?

On the negative side of the ledger, let's consider the following: frequency drift up, down, and around cannot be tolerated; your equipment will have to hold 0.0005 percent frequency stability. The Pre-Prog MC306 and the Motorola T-44 transmitters are rated for this; **however**, junkbox ovens and surplus crystals are **out**; save them for AM use or whatnot. You will have to use the finest crystals available (International or Sentry) and topgrade ovens. (This poses no problem for most of us, as serious FM'ers have always used the best.) Gone will be the days of setting deviation by haphazard rule of thumb; the level will be ±5 kHz—period. You will have to make, scrounge, or buy deviation-measuring equipment because you can no longer use your dev pot for a mike gain control! Seriously though, aren't we truthfully a lot better off without these "disadvantages"?

This is a conversion article for **complete** narrowbanding of the very popular GE Pre-Prog Line MC306. Although the MC306 receiver differs considerably from its Motorola T-44 counterpart, enterprising T-44 owners may use the same general procedure. Unfortunately, the Motorola owner's conversion cost begins at $20, as he must buy a new Permakay

Figure 1

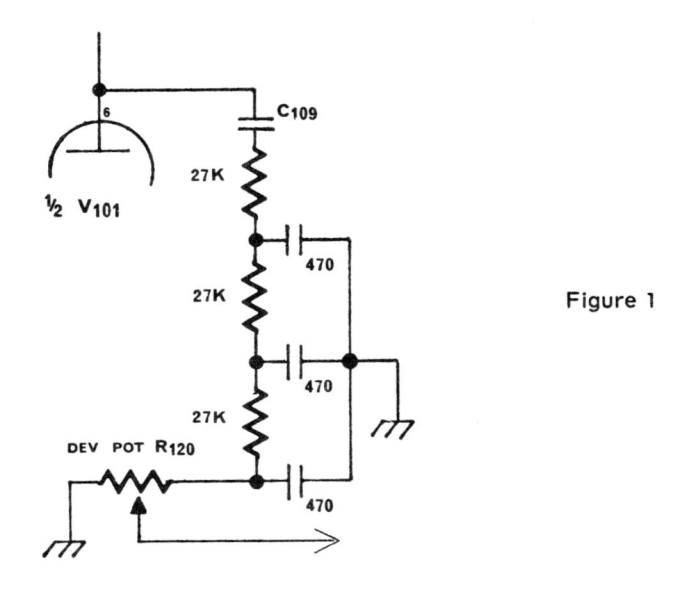

filter; GE owners merely modify and realign the existing six-coil 290 kHz i-f can.

The entire conversion is not difficult, but I strongly suggest that it not be attempted by inexperienced technicians! I-f and discriminator alignment is a serious business and you can make or break a receiver in this one area. You will need a BC-221 or LM-type frequency meter and some type of output attenuation box as well as ordinary shop equipment. Naturally, you should have complete MC306 schematics. In addition to these, you will need a Progress Line manual that gives alignment instructions for the six-coil 290 kHz i-f can used in narrowband 150 MHz receivers.

The conversion may be broken down into four main steps. First, modifying the transmitter modulator and post-limiter filter. Second, increasing the receiver frequency stability to 0.0005 percent. Third, modifying the receiver audio and squelch circuits. And fourth, narrowbanding the six-coil low i-f can. Wherever possible, I will use GE component identification such as "C301," etc.; where necessary, I will refer to schematics shown in this article.

TRANSMITTER NARROWBANDING

Step 1. Remove C103 (0.02 uF) and replace with 0.005 disk.
Step 2. Remove C117 (4700 pF) and replace with 3300 pF.
Step 3. Remove R119 (56K) and in its place install the circuit shown in Fig. 1.

Upon completion of the foregoing, it is only necessary to reset your mod pot to ±5 kHz. Incidentally, the part values given are those specified by GE. Also, your mod circuit is now identical with the later-model GE Progress Line 450 MHz units.

RECEIVER NARROWBANDING
Improving Frequency Stability To 0.0005 Percent

If you have already replaced the original AFC-type first oscillator platter with a Progress Line heated-oven circuit, you may skip this section. If not, convert the original circuit to the Progress Line oscillators as shown in the schematic of Fig. 2. Sketch A of Fig. 2 shows the original oscillator circuit of the Pre-Prog; the modification is shown in sketch B. The sequence is as follows:

Step 1. Remove the AFC-type first oscillator platter and rewire in accordance with sketch B, Fig. 2.

Step 2. Mount an oven socket on the adjacent blank plate and connect its heater to the 6- or 12-volt bus.

Step 3. Remove the following miscellaneous components: S301, R354, C373, C387, R6, R323, and black shielded AFC feed cable.

Your old AFC-type oscillator crystal cannot be used in this Progress Line circuit; it will be necessary to order a new International or Sentry crystal. When ordering the new crystal, specify the following information:

- Receive frequency
- Crystal frequency

$$f_{xtal} = \left(\frac{f_{rcvr} - 48}{36} \right)$$

- GE 4ER26 Progress Line oscillator circuit
- 85°C crystal oven
- F605 holder
- Non-AFC use

Make doubly sure you have a good crystal oven. If you have doubts, check the oven.

Receiver Audio And Squelch Modifications

By narrowbanding your system and running ±5 kHz deviation, you will suffer a serious loss of receiver audio; in addition, due to a lower level of noise into the squelch circuit, it may fail to squelch the receiver. By following the detailed procedure shown here you will eliminate these undesirable effects. Your receiver's final audio output power will be equal to or greater than what you had before.

Figure 2 Sketch A

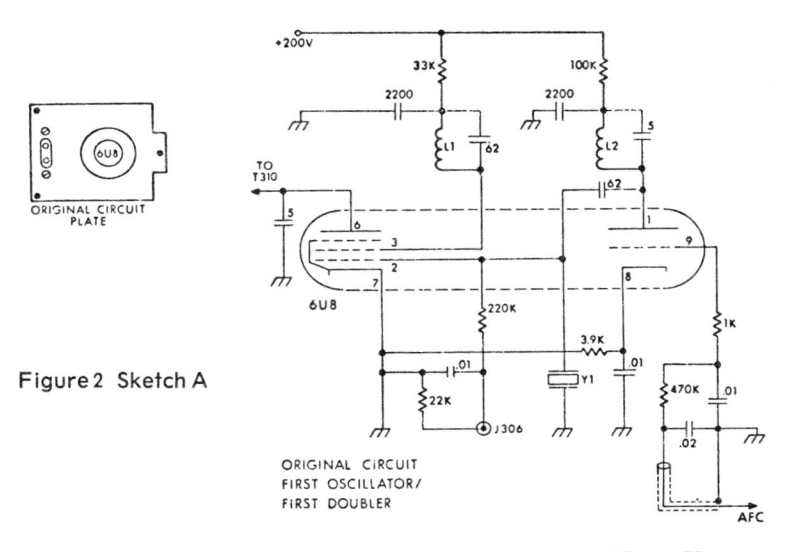

ORIGINAL CIRCUIT PLATE

ORIGINAL CIRCUIT
FIRST OSCILLATOR/
FIRST DOUBLER

Step 1. Remove R353 (100K) and replace with 470K.
Step 2. Remove R352 (100K).
Step 3. Remove C347 (1500 pF).
Step 4. Remove C349 (47 pF) and replace with 470 pF.
Step 5. Connect free end of R360 (220K) to pin 1 of 6AL5 (V313).
Step 6. Remove R361 (100K) and replace with jumper.
Step 7. Remove C357 (1500 pF).
Step 8. Remove R381 (10K) and replace with jumper.
Step 9. Remove C358 (1000 pF) and replace with 0.01 (400V).

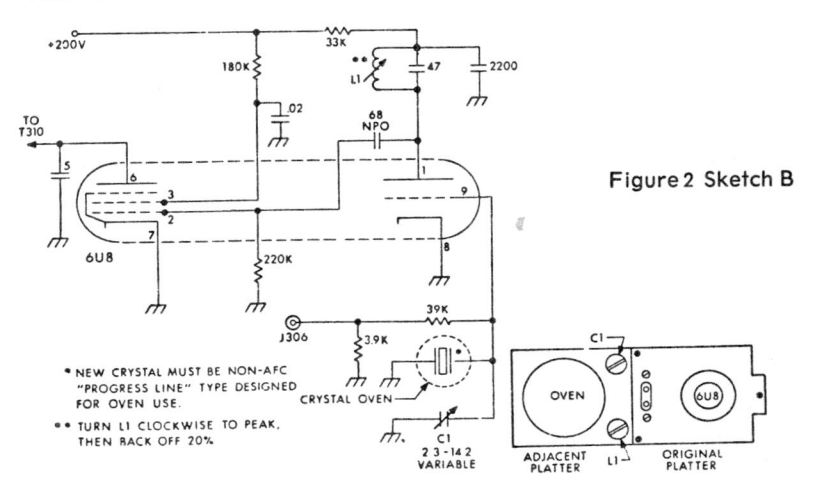

Figure 2 Sketch B

* NEW CRYSTAL MUST BE NON-AFC "PROGRESS LINE" TYPE DESIGNED FOR OVEN USE.

** TURN L1 CLOCKWISE TO PEAK, THEN BACK OFF 20%.

161

Step 10. Add 2500 pF (400V) from pin 1 of 12AX7 (V316) to ground.

Step 11. Replace C361 (4700 pF) with a new component of same value.

Step 12. Remove C359 (0.02 uF).

Step 13. Remove R376 (270 Ω, 1W) and replace it with 270 Ω, 2W.

Step 14. Remove C362 (25 uF, 25V) and replace with a new component of same value.

Step 15. Remove C360 (5 uF) and replace with a new component of same value.

Step 16. Remove R369 (1 meg) and replace with 10 meg.

Step 17. Remove R382 (100K) and replace with 1 meg.

Narrowbanding Low I-F

GE engineers were incredibly farsighted in the design of the Pre-Prog receiver; the six-coil 290 kHz i-f can (T303) is the same unit that is used on later Progress Line units; thus, it is ready-made for narrowband conversion. Either remove T303 and replace it with a new "factory" narrowband one or follow standard GE procedure in narrowing the original. To modify the original, proceed as follows.

Step 1. Remove T303.

Step 2. Carefully remove the three covers shielding the six i-f coils.

Step 3. Clip C3 (7.25 pF) loose from L2.

Step 4. **Very carefully** remove coil assembly L2 and reinstall it in the adjacent "N" hole. Reverse the assembly so that the center hole is again over the test point; **be careful** not to change the position of the coil slug.

Step 5. **Very carefully** remove coil assembly L3 and reinstall in same manner as L2.

Step 6. Remove C3 (7.25 pF).

Step 7. Remove C6 (6.0 pF) and install it in place of C3.

Step 8. **Very carefully** remove coil assembly L5 and reinstall in same manner as L2.

Step 9. Install a 2.0 pF capacitor in place of C6.

Step 10. Replace the three shield covers.

Step 11. Use an ohmmeter to check the center test points on L2 through L5 to ground for possible shorts due to coil handling and coupling capacitor replacement. Normal resistance is a few hundred ohms.

Step 12. Reinstall i-f can T303.

I will not give complete alignment data for T303 here; you should have this information available in a Progress Line 150 MHz narrowband manual. I will, however, give the general procedure to be followed. The first step in the alignment procedure is to check the discriminator transformer (T305). To do this, proceed as follows:

Step 1. Monitor the discriminator test point J310 with an ac VTVM on a low dc scale.

Step 2. Connect an **accurate** 290 kHz generator through a 0.01 capacitor to pin 1 of 6BH6 V310. Saturate the limiters.

Step 3. Adjust the top slug of T305 for zero at J310.

Step 4. Shift the generator frequency **exactly** 5 kHz upward in frequency and carefully note the resultant reading at J310.

Step 5. Shift the generator frequency **exactly** 5 kHz downward from the 290 kHz center and carefully note the resultant reading at J310.

Step 6. Compare the two voltage readings obtained in steps 4 and 5 with one another. If they differ by over 0.1 volt, the discriminator primary should be realigned; if the two readings are essentially the same, stop here and go to the six-coil i-f (T303).

Step 7. If the discriminator primary must be realigned, carefully adjust the bottom slug of T305 a fraction of a turn in one direction and then repeat steps 2 through 6. Continue shifting the bottom slug of T305 in small increments one way or another until the readings at J310 are essentially equal at 5 kHz above and 5 kHz below the center frequency.

To properly align the six-coil 290 kHz i-f can, connect your generator through a 0.01 capacitor to pin 1 of 12AT7 (V308). Monitor the discriminator test point (J310) with a VTVM or VOM. Monitor the limiter test point (J309) with a **VTVM** only. Adjust your generator for **exactly** 290 kHz. If your discriminator alignment was correct, you should read dead zero at J310. You are now set to follow the GE resistor loading method of alignment.

Your output attenuator will be invaluable as you adjust coils L1-L6. Simply keep increasing the attenuation as the limiter reading rises during alignment. The foregoing procedure may sound somewhat hairy, but after your first one you should be able to go through the necessary alignment in a few minutes! Be patient the first time as you get your feet wet.

Commercial shops should carefully note the following. As of November 1968, the GE MC306 or 4ES14A1 was scheduled to lose "type acceptance" completely. Although the transmitter postlimiter modifications specified herein transform the unit into a true narrowband transmitter, there is no indication at this time that the FCC will "buy" this and permit the unit to retain type approval.

A New Lease on Life for GE 450 Pre-Prog Receivers

by Jim Lev K6DGX

Here is a worthwhile modification that will increase the MTBF (a military abbreviation for "mean time between failure") of your "Pre-Prog" mountaintop receiver or mobile unit. Other than the standard afc first-oscillator platter, the 6AM4 rf stage is the second weak link in an otherwise fine receiver; the 6AM4 is the most often replaced tube in the Pre-Prog receiver.

Amperex makes a tube, the ECC88 / 6DL4, that has proved itself in hundreds of these receivers. While it offers no really noticeable improvement in receiver sensitivity, it doesn't degrade performance by one iota, either. What it **does** do is last and last. Who doesn't want reliability?

Cost is no problem, either. According to a recent price list, the 6AM4 sells for $2.72 while the Amperex ECC88 / 6DL4 goes for $2.60. Once you have accomplished the simple modification on your first receiver, you'll be able to perform the necessary socket wiring changes in ten minutes.

I claim no credit for originating this innovation; the originator's name is—as the proverbial saying goes—"lost in antiquity."

This modification is being used with great success by many service shops. In the Los Angeles area, it is a standard modification for Mobile Communications Service, Chapman TV & Electronics, and Mobilfone.

The Pre-Prog receivers in my own W6FHF remote system at Blue Ridge Summit are also modified with the Amperex tube.

The necessary tube socket wiring changes should be quite evident after a careful study of the pictorial wiring diagram. A 2.2 pF NPO capacitor is added from the 6DL4 plate to ground because of a slight difference in interelectrode capacitance and to facilitate proper rf stage tuning in the 440-450 MHz region. Naturally, the lead length should be as short as physically possible. Filament choke L309, which normally has the longest leads, is removed from pin 7 and reused as shown at pin 5.

UNDER CHASSIS VIEW

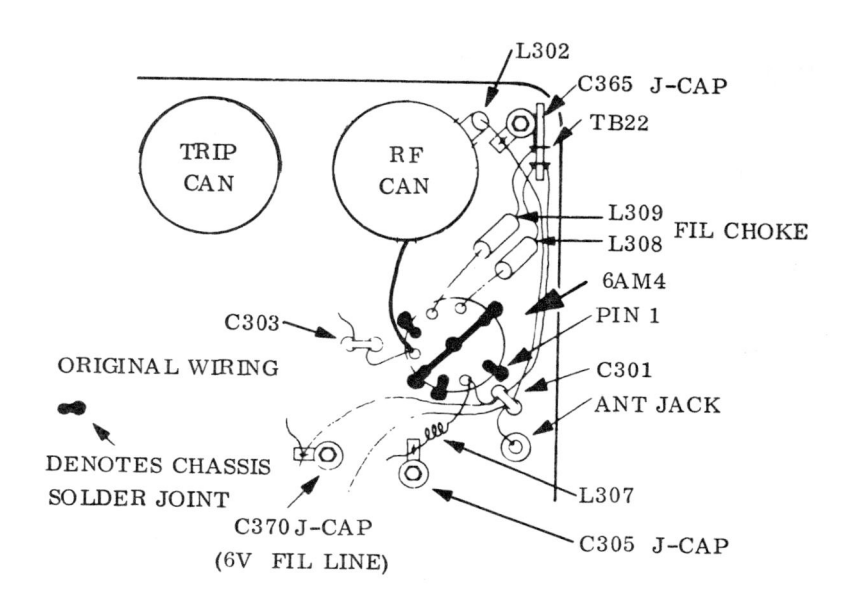

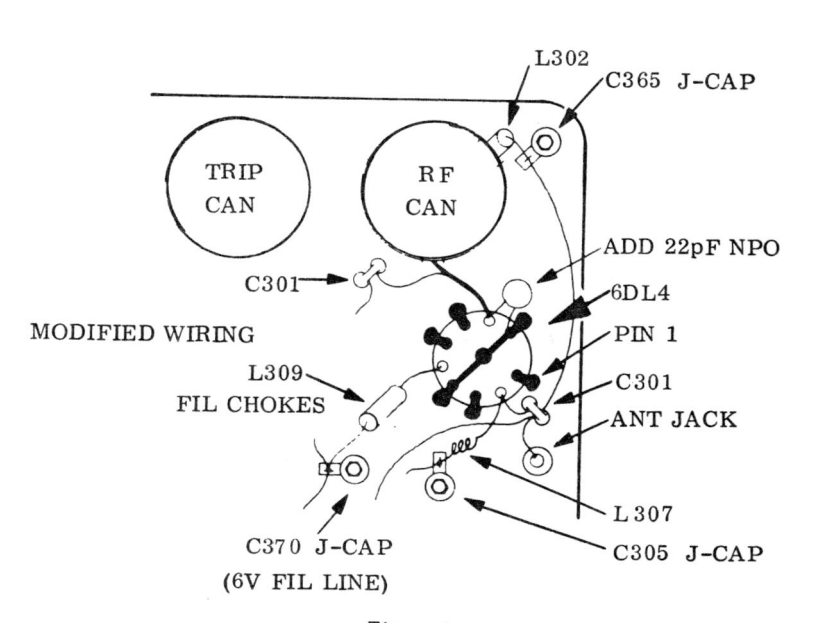

Figure 1

A word of caution is in order: When relocating the plate bus strap from pin 5 over to pin 8, be very careful not to use too much "vigor"; this bus lead can be broken or shorted inside the rf shield very easily. And remember, **always** check the plate pin to ground with an ohmmeter before applying power: You want an improved receiver, **NOT** a smoke generator!

An AC Supply for the Motorola H23 Handie-Talkie

by Richard Thomas W8VJC

The Motorola H23BAM Handie-Talkie, with the all-transistor receiver, comes with an 8-pin Jones plug installed so the unit may be operated from an external 6- or 12-volt battery supply. Provisions are built-in to trickle-charge the internal ni-cad 6-volt battery at a 70 mA rate while the H23 is being used on an external 12-volt battery. With proper connections to the external power supply plug, a regulated 12-volt dc power supply will operate the transceiver and charge the Ni-Cad battery from the 117-volt ac line, allowing you to use the Handie-Talkie around the ham shack and still keep the battery charged for portable operation.

The regulated power supply shown will give close to 12 volts out, with current variable from a trickle charge of 60 mA to full transmit and fast charge of 1.35 amps. Fast charge is available by switching a No. 44 dial lamp connected through a diode to pin 1 of P201. This adds 230 mA of current to the 70 mA of trickle-charge current built into the Handie-Talkie to make up a total of 300 mA for fast charge. The diode in series with the fast-charge switch is there so that the ni-cad battery will not discharge back through the receiver if the power supply is switched off. The No. 44 lamps acts as a pilot and also as a fuse if the battery is internally shorted or completely discharged.

I mounted all the diodes, capacitors, and small parts on an old piece of printed circuit board. The power transformer, switches, and pilot lamps are on a 3½ x 4½ x 1 inch chassis. The circuit board is under the chassis. If you keep most of your transmitting times short, the transistor may be mounted on the side of the chassis with a power transistor mounting kit, using the chassis as a heatsink. In the transmitting mode, the 2N255A transistor must dissipate over 6 watts of power and will get warm. If it becomes hot while making many transmissions, you may want to use a commercial power transistor heatsink with a higher-rated PNP power transistor in the power supply.

The ni-cad battery supplied in the H23BAM Handie-Talkie is a Motorola NLN 6134A battery rated at 6 volts, 4 amp-

Figure 1

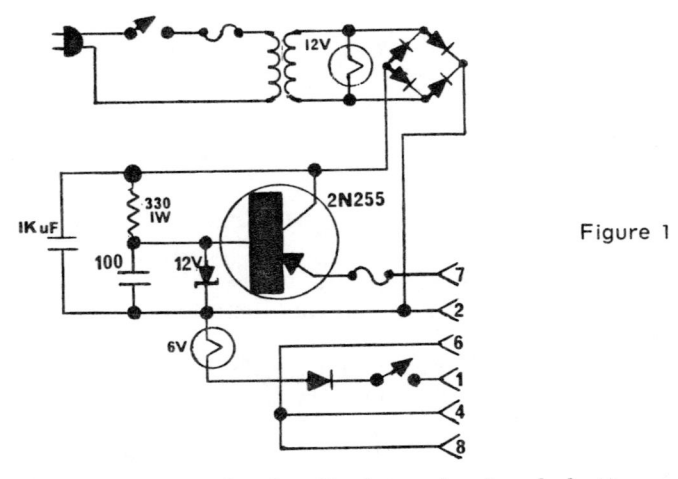

hours. A completely discharged ni-cad battery (battery voltage below 3 volts) requires more than 15 hours at 10 percent of the milliampere-hour rating to 24 hours of fast charge to restore it to full charge. Continuous trickle-charging of a ni-cad battery at 50 or 60 mA will not harm the battery. If a fully charged ni-cad battery stands idle for 6 weeks, it will discharge to 75 percent of full charge. The NLN 6134A battery is designed to operate the H23BAM transceiver for 8 hours with 10 percent transmitting time on a full charge. After that the battery voltage will have dropped to about 5 volts. Then 8 to 12 hours of fast charging should restore the battery to full charge. (Fast charging is defined as a charge current equal to more than one-tenth the milliampere-hour rating but less than 20 percent of this figure.)

Plug-In FET Circuits Replace Vacuum Tubes

by P. J. Ferrell

Motorola introduced the first commercially available transistorized portable FM transceivers some time around 1956. These beautiful little units are now widely available to amateurs, but they come in a bewildering assortment of type numbers. For example, a P33-4 is a single-frequency transceiver for the 144-174 MHz region with 7 watts output, microphone, speaker, and rechargeable nickel-cadmium battery; an H23AAC-310AH is a high-band split-channel one-watt unit with handset and extra-duty dry battery.

The year 1956 was such a short time ago that it is sometimes startling to remember that available transistors at that time would not oscillate above about 1 MHz—and anything above that was vacuum tube country. So it was that the local oscillator and the two first i-f's of the Motorola HT receivers used vacuum tubes. Later units were completely transistorized, and a conversion kit is still available from Motorola (NED6004A, $74.00) which updates the early receivers to the fully transistorized configuration. If your unit has the late-model receiver, go immediately to some other article, because the remainder of this one will just make you wish you hadn't splurged on the nonhybridized vintage.

Using inexpensive N-channel field-effect transistors, the early receivers can be readily converted to fully solid-state operation, and these modified receivers will perform rings around the newer units (which use bipolar transistors rather than the hotter FETs). The necessary modifications will cost about $7, and the work can be finished in little more than an hour.

The FET cascode shown in Fig. 1 is generally useful as a pentode vacuum tube replacement. The transistors specified are readily available either locally or from the larger mail-order supply outlets. Other N-channel FETs, such as 2N3823 and those of the Motorola MPF series, will work equally well. In this application, the supply potential of 50 volts is just right as the two FETs are in series for dc and each gets about 27 volts from drain to source. Resistor R3 determines the current drawn by the series transistor pair. The current in milliam-

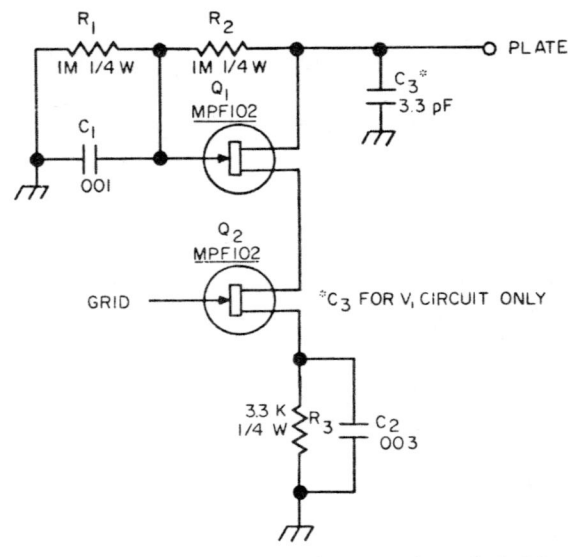

Fig. 1. FET cascode arrangement replaces each pentode i-f amplifier in the H23 hybrid receiver. (Note that C3 is required for the first i-f amplifier, though it is not used in the second.)

peres is approximately 2000 / R3. More than adequate gain is obtained at a current of 600 microamperes, and with the higher gains obtained at higher currents, stability problems can arise.

The output capacitance of this circuit is negligible, so that C3 is required to correctly tune the output of the V1 tube replacement. There is sufficient capacity in the output circuit of the second stage (V2) so that a C3 equivalent is not required.

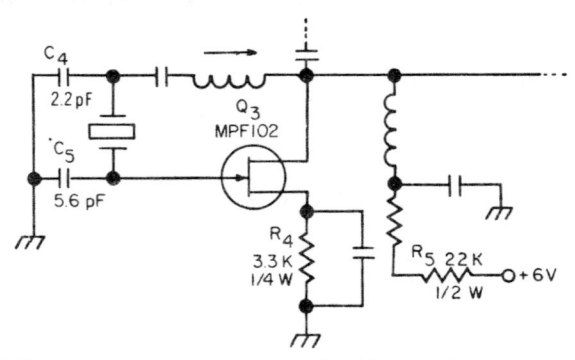

Fig. 2. FET replaces vacuum tube in the H23 oscillator to complete the transistorization operation. Components not labeled in the sketch are those components that are already part of the existing oscillator circuit.

Vacuum tube V3 is triode-connected, so a single FET is used. Capacitors C4 and C5 return the crystal operating frequency to that of the original tube circuit. Resistor R5 drops the 60-volt B+ to a level that is safe for the FET. (See Fig. 2.)

As a last touch, replace the first crystal-mixer diode CR1 (it will be a 1N72 or a 1N147A) with an HP 5082-2800 hot carrier diode. This will only cost a buck, and the expense is worthwhile. My own wideband H23AAM measures better than 0.4 microvolt sensitivity at 20 dB of quieting, and adjacent-channel problems due to cross modulation have disappeared.

6 Freq. Conversion:
80D & 140D Transmitters
by Charles Copp W2ZSD

In an effort to obtain the maximum number of FM channels per transmitter, it is necessary to package as many oscillators as possible in the available space. For the Motorola 80D and 140D transmitters, the three oscillator decks can be removed and a single plate made to cover the existing opening. Without too much difficulty, it is possible to put four oscillators on this plate. See Fig. 1 for the layout of the four-frequency plate.

When I decided that a fifth frequency was desirable, I was in trouble. It now became a matter of whether to mount a second transmitter strip in my already crowded cabinet or to get more oscillators into the existing space available (2¾ x 4⅜"). After juggling components around, a six-frequency oscillator deck was decided upon and assembled with all components fitting neatly into their prescribed places. However, as can be seen from the photos, certain sacrifices and modifications have been made.

First of all, ovens are out of the question for this design, and second, tube shields may have to be left off. However, with careful tube socket placement and the use of the black slipover

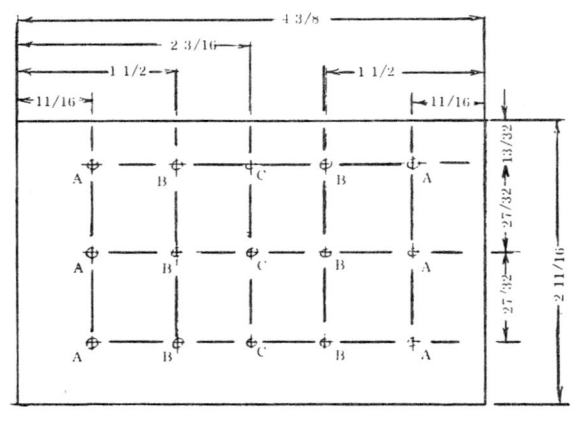

Figure 2

HOLES:
"A" 5/8" Dia. (for tube socket)
"B" Mounting hole for trimmer
"C" Center of crystal socket cluster

tube shields, it is possible to shield the tubes, but it isn't necessary. The dual crystal sockets accommodate two crystals back to back. The air trimmer capacitor used by Motorola must give way to a piston capacitor. The ones I used were JFD MC604 with a tuning range of 1 to 42 pF. The exact tuning range is not important since the 10 pF capacitor across the trimmer can be made larger or smaller if there is not enough tuning range to put the crystal on frequency. Suitable piston capacitors can usually be found in the surplus stores.

For those who insist on using ovens it is possible to mount six dual sockets between the tube sockets in place of the three dual sockets. The piston trimmers are then replaced with ceramic trimmers mounted beneath the oscillator deck. Frequency adjustments will have to be done from below the chassis, but this should present no great problem. Ceramic trimmers can even be substituted for the piston trimmers in the no-oven design if pistons are not available.

Figures 1 and 2 show the layout of the four- and the six-frequency decks, respectively. A schematic has not been included since the circuit is an exact duplication of the Motorola P865 oscillator deck circuit.

The basic transmitter must also have some wiring changes in order to accept the six-frequency deck. There are twelve terminals available for the original three oscillators, and ten of these terminals will be used.

The plates of each group of three oscillators are wired together and brought out to the nearest terminal, keeping the leads as short as possible. The remaining leads are not critical and can be wired in any convenient arrangement. It is desirable to keep the transmitter strip capable of being plugged in, so five of the oscillators will be wired to the power socket (pins 1, 2, 3, 9, and 11) and the remaining oscillator will be wired to pin 9 on the meter socket.

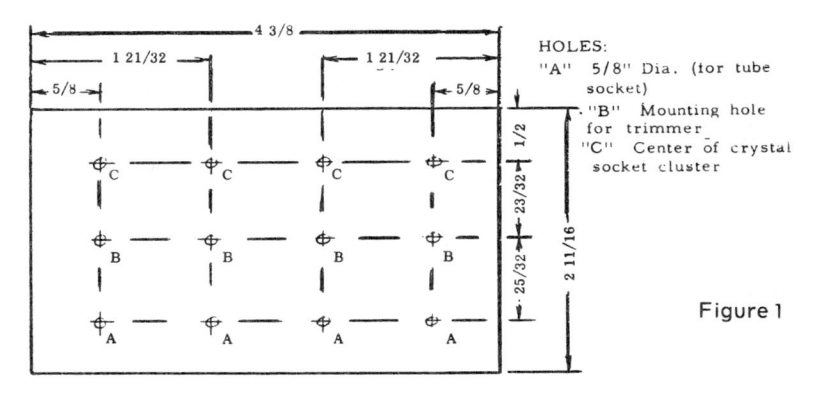

HOLES:
"A" 5/8" Dia. (for tube socket)
"B" Mounting hole for trimmer
"C" Center of crystal socket cluster

Figure 1

NOTE: Use the power socket connections to control the two frequencies furthest apart so that they can be switched back and forth while tuning up with a test meter inserted in meter socket. Also, since there are jumpers between some of the terminals on the oscillator deck terminal strip, the unneeded jumpers will have to be removed.

A little effort was involved in getting the completed deck into the transmitter strip, but it does fit. It might be necessary to notch out slightly for the crystal or tube sockets. However, the extra work is worthwhile in order to have six transmit frequencies available using a single transmitter strip.

I would like to thank John Olsen (W2WJS) for supplying the photographs of the six-frequency oscillator deck.

4-Frequency Conversion for the 450 Pre-Prog
by Jim Lev K6DGX

The GE Pre-Progress Line 450 MHz mobile unit is the most versatile piece of equipment available to the FM amateur today. Unlike its nearest rival, the Motorola T44, it is blessed with a fair amount of open chassis space that may be used to great advantage. (The T44 has some open underchassis space, but only on the power supply strip.) The four-frequency circuit diagram illustrated here is a composite of existing GE circuitry that has been modified to permit use in the Pre-Prog receiver.

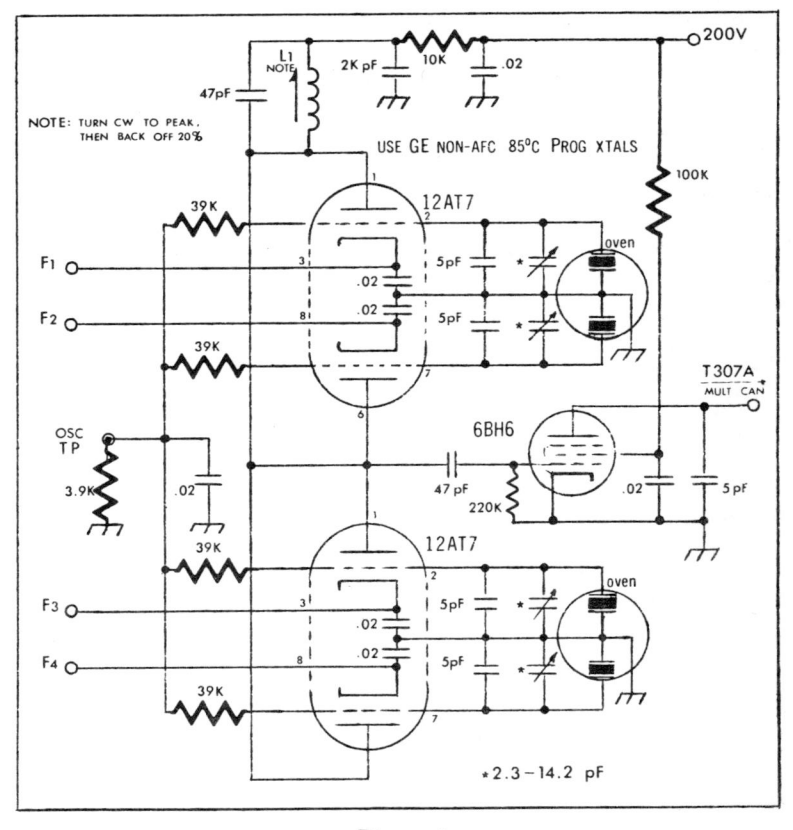

Figure 1

In addition to the four-frequency receive capability, there is also a decided advantage to be realized by eliminating the existing AFC oscillator platter due to its tendency to excessive frequency drift.

To begin the conversion remove the existing AFC first-oscillator platter from the receiver. Remove the three adjacent blank oscillator platters, then remove the AFC feedline (black shielded cable running towards rear of receiver strip). A metal plate should now be cut to fit into this opening; a little metalwork with a nibbler tool is in order to make sure cross pieces obstructing this opening are cleared away. Coil L1 in the schematic is the same coil that is on the AFC-type oscillator platter; it must be carefully removed and remounted on the new plate; examine the coil slug carefully to be sure it has not been "diked"; this is a favorite stunt of some commercial technicians that is sometimes used to "stretch" receiver frequency without having to buy a new crystal; remove the present capacitor across L1 and solder a 47 pF in its place.

It will of course be necessary to employ dual crystal ovens; for best results it is suggested that either GE, ITT, or RCA 12-volt types be used. These are available at $3 each from Mann Communications, 18669 Ventura Blvd, Tarzana, California.

Your present AFC-type oscillator crystal **may not** be used as the correlation is different; it will be necessary to buy new crystals of the GE "Progress Line" type that are available from Sentry Manufacturing Co. When ordering, specify that the crystals are to be used in a GE Progress Line Model No. 4ER26 **NON-AFC** receiver; holder type F-605 with an oven temperature of 85°C.

Converting the Motorola 41V

by Don Milbury W6YAN

Setting up the 41V to operate on two meters is very simple. After a few preliminary value adjustments, the tuneup procedure is perfectly straightforward. This deals primarily with changing a 6V unit to 12V; but, in the interest of completeness, here is the frequency conversion information:
    1. Add 2-5 pF from pin 1 to pin 3 on L1, L2, L3, L4, and L5 of receiver.
    2. Add 4 or 5 pF from pin 1 to pin 3 on L7, L7A, and L8 of receiver.

Crystal data:
    Transmit—Motorola Type R03, 85ºC oven. Specify operating frequency; Sentry will correlate.
    Receive—Motorola Type R21, 85ºC oven. Specify operating frequency; Sentry will correlate.
Sentry address is: 1634 Linwood Blvd, Oklahoma City, Ok 73106.

6 to 12V CONVERSION, TRANSMITTER

a. Doubler-Driver (V106)
    (1) Disconnect and remove the jumper between pins 7 & 8.
    (2) Disconnect the brown-white lead from pin 2 and connect it to pin 7. This connects tubes V106 and V107 in series.

b. Tube V105: 3rd Doubler (25-50 MC) 2nd Doubler (152-174 MC)
    (1) Disconnect and remove the ground lead from pin 3.
    (2) Disconnect and remove the jumper between pins 2 & 3.
    (3) Connect a jumper between pin 2 and the center shield.
    (4) Disconnect the brown-white to pin 3. This connects tubes V101 and V105 in series.

c. Audio Amplifier (V108)
    (1) Disconnect and remove the ground lead from pin 9.
    (2) Remove the jumper between pin 9 and the shield.
    (3) Connect a jumper between the center shield and ground.

(4) At the V108 tube socket, disconnect the brown lead (running from tube V109) from pin 4 or pin 5, depending upon which is used, and connect it to pin 9. This connects tubes V108 and V109 in series.

d. Tubes V102 (Modular), V103 (Buffer and 1st Doubler), and V104 (2nd Doubler—25 to 50 MHz and Tripler—152 to 174 MHz)

(1) Replace the three type 6AU6 tubes (V102, V103, and V104) with type 12AU6 tubes.

e. Antenna Relay

(1) Remove the three screws that hold the antenna relay cover to the chassis. Unsolder the relay cover from the shield around the rf section.

(2) Remove the solid bus which connects the relay coil to ground.

(3) Connect the brown-white lead between the relay coil lug, from which the ground bus was just removed, and pin 2 of tube V106.

(4) Replace the antenna relay assembly on the chassis by means of the three screws. Resolder the relay cover to the rf shield. This connects the relay in series with the transmit-receive relay on the power supply chassis.

f. Crystal Socket (Single-Frequency Models Only)

(1) Use items 4 through 8 and mount the two 15-ohm resistors near the crystal socket by using existing holes in the chassis. Place one lockwasher, one eyelet, and one fiber washer at each end of the resistor. Connect the resistors in parallel.

(2) Disconnect and remove the ground connection from the crystal socket.

(3) Connect one end of the paralleled 15-ohm resistors to the crystal socket terminal from which the ground connection was just removed.

(4) Connect the other end of the paralleled resistors to ground.

g. No. 2 Oscillator V201 (Two-Frequency Models Only)

(1) Disconnect and remove the jumper between pins 2 and 3 of the V201 tube socket.

(2) Connect the 39-ohm resistor between pin 3 of tube V201 and the grounded lug on the crystal socket (X202). This connects the filament of tube V201 in series with the 39-ohm resistor across the 12V source.

h. Crystal Assemblies (Two-Frequency Models Only)

(1) Make a note of the frequency of each crystal and remove the two crystal assemblies from their sockets. The

serted in the new heater and base assembly as outlined in the following steps. Use care so that 6 and 12V assemblies are not mixed.

(2) Remove the housing cover by releasing the ring clamp.

(3) Remove crystal from the 6V base assembly and install in 12V base assembly.

NOTE: The letters A and B are stamped on the bottom of the base assembly. Plug the crystal into the socket on the A side.

(4) Place the spacer plate between the crystal and the heater element and replace the cover. Make sure that the polarity of the cover and base agree.

(5) Insert the crystal assembly in the proper socket.

6 TO 12V CONVERSION POWER SUPPLY

a. Vibrator Connections

(1) Disconnect the shielded lead from the two 2.9H choke coils located near the center of the chassis (Fig. 1).

(2) Disconnect the other ends of the two chokes from the two 0.5 mF capacitors (C4 and C5). These two choke coils are not used in the 12V circuit.

(3) Disconnect the black-yellow transformer lead from capacitor C4 (0.5) and disconnect the red-yellow transformer lead from capacitor C5 (0.5). Tape each lead separately and dress it out of the way.

(4) Disconnect the sleeve-covered lead from pin 3 of the vibrator socket. Use this lead to connect capacitors C4 and C5 in parallel. Do not solder the connections to C4 and C5 until completing step 15.

(5) Disconnect the black, yellow, and red leads from pins 1, 2, and 4 of the vibrator socket.

(6) Remove the lead between vibrator socket pin 6 and ground.

(7) Remove the 100-ohm resistors connected between vibrator socket pins 1, 2, 4, and 5 and ground.

(8) Connect the red and black leads to the ungrounded terminal of C4 or C5 (changed to C104 and C105 on schematic).

(9) Connect pins 2 and 4 of the vibrator socket and then ground pin 4 at the ground lance near pin 5.

(10) Connect a 270-ohm resistor from vibrator socket pin 1 to ground.

(11) Connect a 270-ohm resistor from vibrator socket pin 5 to ground.

(12) Connect a 7.5-ohm resistor from vibrator socket pin 6 to ground.

Figure 1

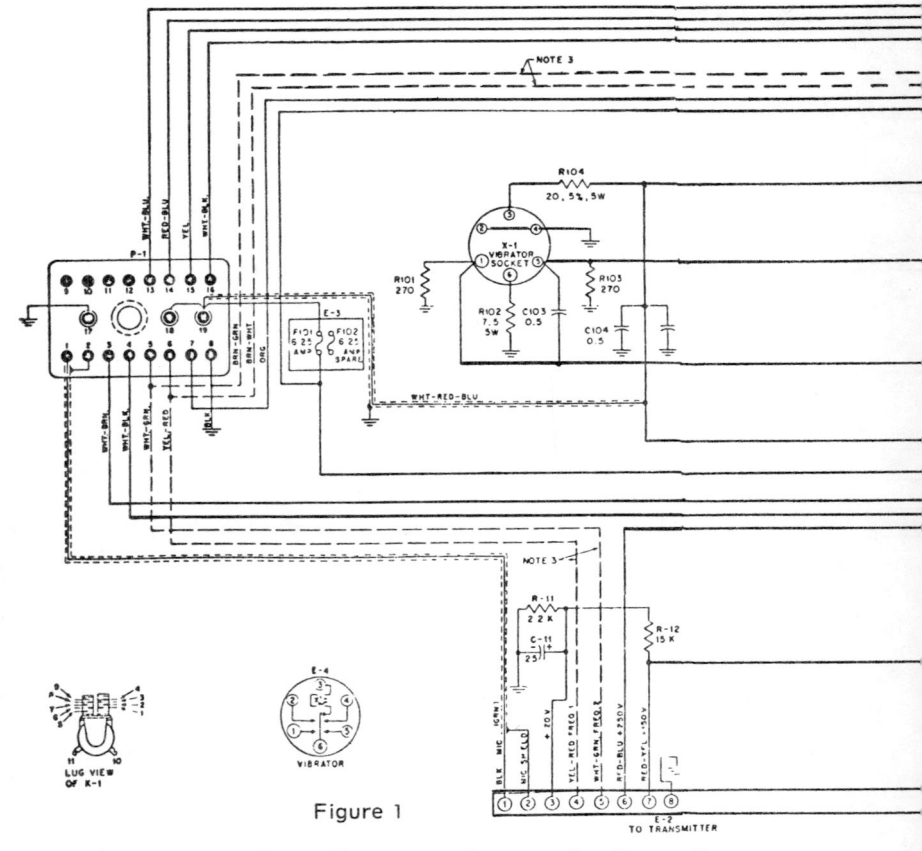

(13) Connect the yellow transformer lead to vibrator socket pin 1.

(14) Connect a 20-ohm resistor from vibrator socket pin 3 to the ungrounded terminal of capacitor C4.

(15) Connect the shielded lead, which was disconnected from the 2.9H choke coil in step 1, to the ungrounded terminal of capacitor C5.

b. Push-to-Talk Relay Modifications

(1) Remove the white-black lead connected between terminal 8 of relay K1 and terminal 14 of transmitter terminal strip E2.

(2) Remove the white-black lead which connects between terminal 14 of terminal strip E2 and pin 4 of power plug P1.

(3) Connect a black-white lead from terminal 8 of relay K1 to terminal 4 of the power plug.

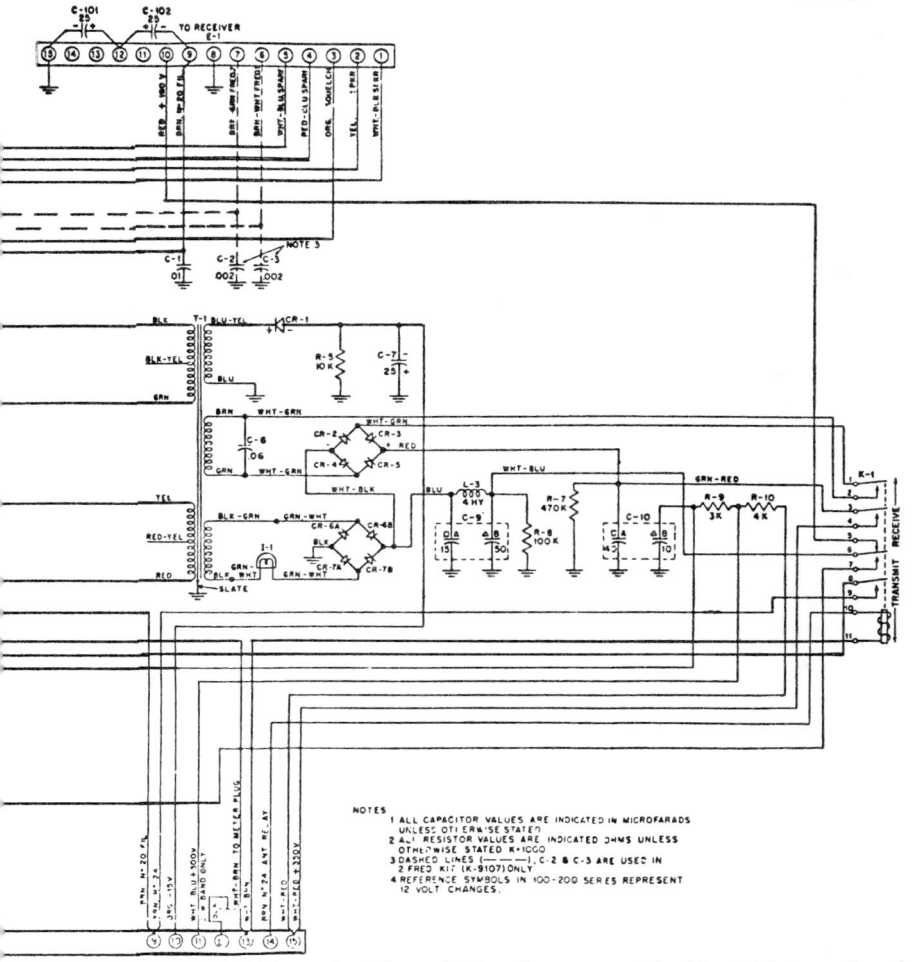

(4) At terminal board E1, disconnect the No. 24 brown lead (running from the relay coil lug) from terminal 9. Reconnect this end of the lead to terminal 14 of terminal board E2.

d. Fuses

(1) Remove the two 15-ampere fuses from the fuseholder.

(2) Paint out or use tape to mask over the markings on the fuseholder. It is recommended that 6.25 amp now be marked on the fuseholder to insure that oversize fuses are not used.

(3) Place two 6.25-ampere fuses in the fuseholder.

e. Terminal Strip E1 Modifications

(1) Connect a 25 mF capacitor from terminal 9 to terminal 12 of receiver terminal strip. Connect the negative side of the capacitor to terminal 9.

(2) Connect a 25 mF capacitor from terminal 12 to terminal 15 of the receiver terminal strip. Connect the negative side of the capacitor to terminal 15.

Converting the Handie-Talkie

by Bob Lyon WA6DTG

The low prices and ready availability of Motorola tube-type Handie-Talkies make them ideal units for ham use. The 150 MHz quarter-watt unit (FHTRU-1 etc.) is particularly desirable for two-meter use because it is small and lightweight, and it can be converted with a minimum of modification.

For serious AREC work, the Handie-Talkie is almost indispensable, because it allows the operator to get right in and move to where the action is. Emergencies know no limitations, and can happen where there is no power available. Or they can happen to the power. So, it pays to have the capability of maximum transportability in preparation for any type of unforeseen situation.

The major nuisance with Handie-Talkies is the battery pack aspect: Batteries are expensive and cannot last long under the drain of heavy or repeated use. One way to preserve the batteries and squeeze additional performance from the hand-carried radio is to build up an ac supply capable of adequately powering the unit when it is not in use in the field. This ac supply can also be used to provide a limited charge for the B batteries during periods of nonuse.

The radio portion of the Motorola Handie-Talkie is built on a long subchassis that mounts on the larger battery box, serving as the base. The battery container portion of the unit makes an ideal chassis for an ac supply. If an extra one of these can be obtained, by all means use it for this purpose.

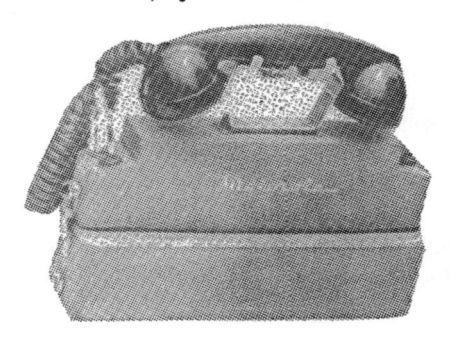

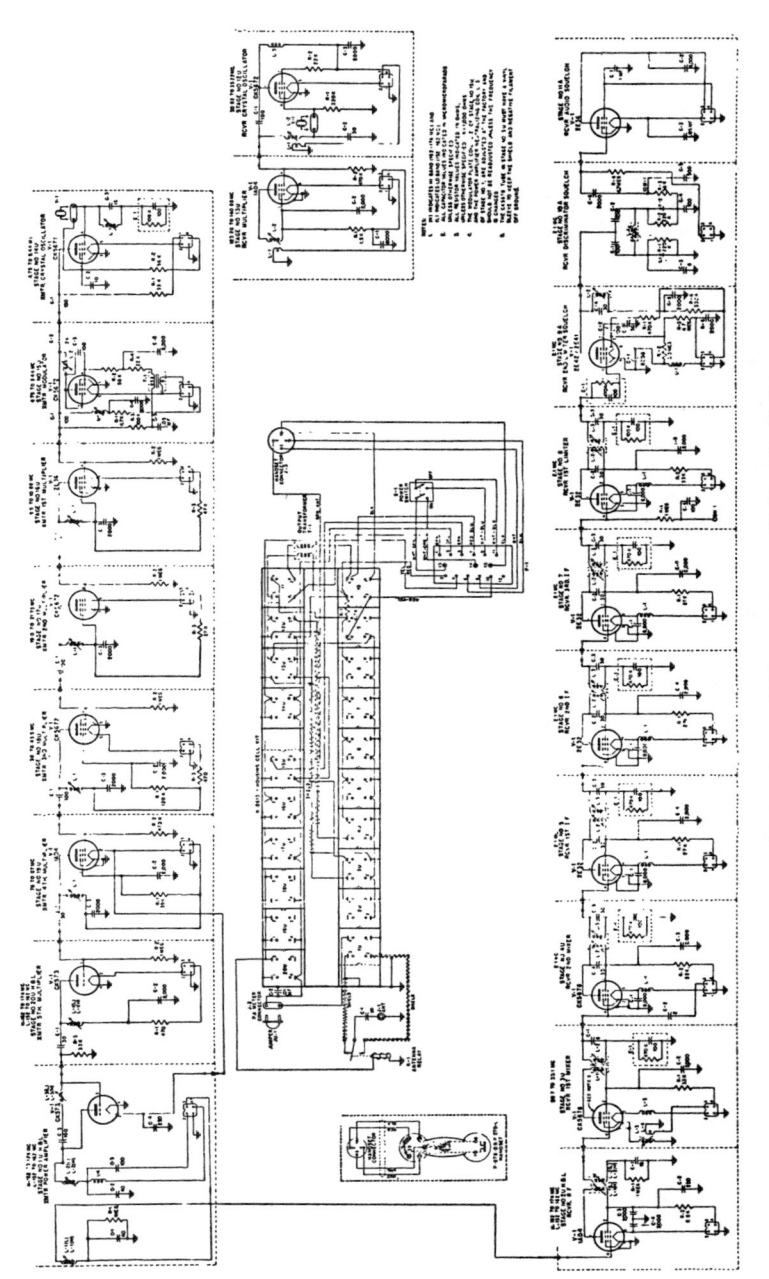

Fig. 1. Motorola 150 MHz quarter-watt Handie-Talkie.

Fig. 2. The bottom section makes an ideal chassis for an ac supply.

The following modifications will bring about satisfactory operation; however, it must be noted that great care must be taken to precisely follow directions because this unit is very responsive to any frequency deviation throughout. An electronic voltmeter is a must (VTVM or equivalent)!

Use a 1 megohm resistor on the tip of the voltmeter probe. Allow no more than a quarter-inch of lead to be exposed on the probe end of the resistor.

A communications receiver can be used to check peaking on the oscillator adjustment of the receiver and for the oscillator and multiplier adjustments of the transmitter up to 30 MHz. The S-meter will serve as a relative scale for this tuneup. A conventional two-meter FM base station receiver set to monitor first limiter current will aid in peaking the rest of the transmitter. Be sure to remove the antenna from the base station receiver and use a 4- to 6-inch piece of wire in the coaxial receptacle to attenuate the signal (otherwise the tuning will be too broad).

For receiver peaking, use the electronic voltmeter at the connecting terminals between 7U and 8U for all stages ahead

Fig. 3. Photo shows complete strip with modules. The transmit and receive oscillator modules have been slid forward for easy identification.

of this. Use terminals connecting 9U and 10U to peak 9U. Then check 10U and 7U for best quieting and noise on frequency. The transmitter is straightforward and should give no trouble at all in tuning.

TRANSMITTER SECTION

Oscillator, 14U (Print 316)
1. Connect 10 pF across crystal socket terminals.
2. Connect 20 pF across the inductance.

Driver, 20U (Print 319)
Connect 1 pF across the inductance.

Final, 1UL (Print 305)
Connect 1 pF across the **transmitter** inductance (7 turns). Do not confuse the inductance with the slug-tuned neutralizing coil on top of assembly. Also, be sure to find proper inductance; there are two: one for transmitter, one for receiver. Check print to be safe.

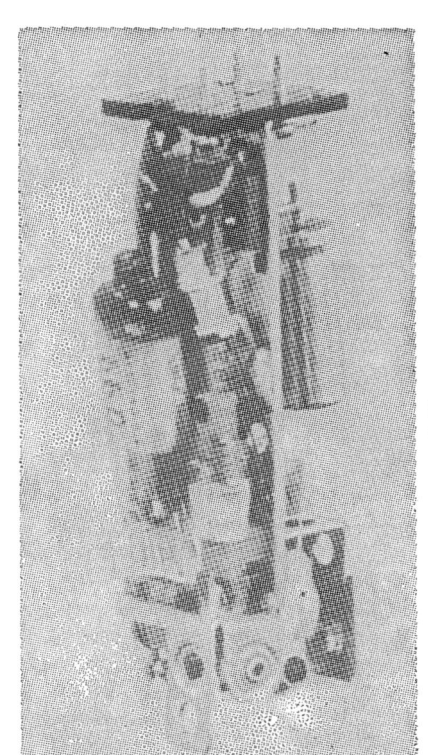

Fig. 4. Transmitter oscillator module.

Modulator, 1UL (Print 317)
 Modulator will not peak out very well; the peak is low, but should be acceptable if tuning is anywhere in the general range.

RECEIVER
Antenna Stage, 1UL (Print 305)
 1. Add a gimmick (2 inches of hookup wire) to high-frequency coil at button connector at top of assembly (connecting 1U and 2U).
 2. Dress down into module through spare hole in top of assembly.

RF Stage, 2UL (Print 300)
 1. Add a gimmick to pin 1 of tube socket.
 2. Dress down into corner of assembly behind primary coil connected to pin 1.

Oscillator
 Peak oscillator by watching signal on S-meter of communications receiver, then back off slightly on oscillator's most stable slope.

First Mixer, 3UL (Print 301)
 1. Add a 1.5-inch gimmick to coil going to pins 1 and 2 of tube.
 2. Dress leads down inside module behind coil affected.
 NOTE: The antenna input coil for the receiver is in the same module (1UL) as the transmitter output tank coil.

SERVICE TIPS & INFO

The Fine Art of Receiver Alignment

Aligning an FM receiver is a great deal more complex than getting the oscillator on frequency and peaking the various stages to an on-channel signal. Yet this is precisely what many amateurs—and, unfortunately, many commercial service technicians—actually do.

When an FM receiver is tuned up using this procedure—we'll call it the "tweak" method—the technician is making a number of raw assumptions which may or may not be valid. First, he's assuming that the low i-f's and the discriminator are correctly aligned to their respective frequencies. Second, he's assuming that the sealed bandpass filter is properly tuned to its design frequency. The latter can generally be a safe assumption, even though it is not uncommon for these filters to change or shift a bit in frequency as a result of excessive vibration or shock or other abuse.

What happens when the tweak method is used for tuneup? It is, admittedly, a quick-and-dirty process by which a receiver can be made to operate. The brutal truth, however, is that the primary receiver qualities of selectivity, sensitivity, and stability are interrelated. The tweak method is an optimum compromise of the three based on the initial setting of the second converter and the low i-f circuits.

Selectivity (and gain, of course) is broadly determined by the number and state of the tuned circuits in the receiver chain, from the antenna, itself, to the discriminator. Each frequency-sensitive element adds somewhat to the selectivity and affords at least some degree of gain. An important point is that each of these elements must be centered on the frequency of operation. To assure proper tuneup of a receiver, the selective circuits (r-f, high i-f, and low i-f) must be aligned so that desired signals can pass through the center of each selectivity curve. Equally important, the configurations of the various curves must conform to their design shapes. The proper combination of these shapes will yield an overall response curve as shown in Fig. 1.

Figure 2 shows how the ideal composite selectivity curve is obtained. In the sketch, the center line represents an in-

Figure 1

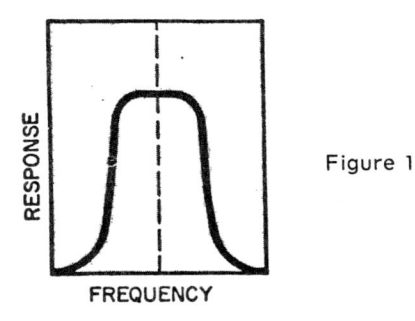

coming on-channel signal. The flowing V at the top is the discriminator slot. The low i-f passband is the steep-sided peak with the broad plateau across the operating frequency. The broader curve with the sharper arc in the frequency range of interest is the selectivity curve of the high i-f. The rf amplifier and antenna are shown as low broad arcs. The curves are plotted as bandwidth (horizontal) versus gain (vertical).

At this point, it would be wise to say that i-f alignment usually isn't necessary unless:
- A component has been replaced in an i-f filter;
- The circuits have been subjected to tweaking without proper test equipment.

Unfortunately, the latter is more usually the case with amateur FM equipment. No amateur should ever try to tune up an FM receiver unless he has a schematic diagram of his equipment so he will know where NOT to tweak. Even in

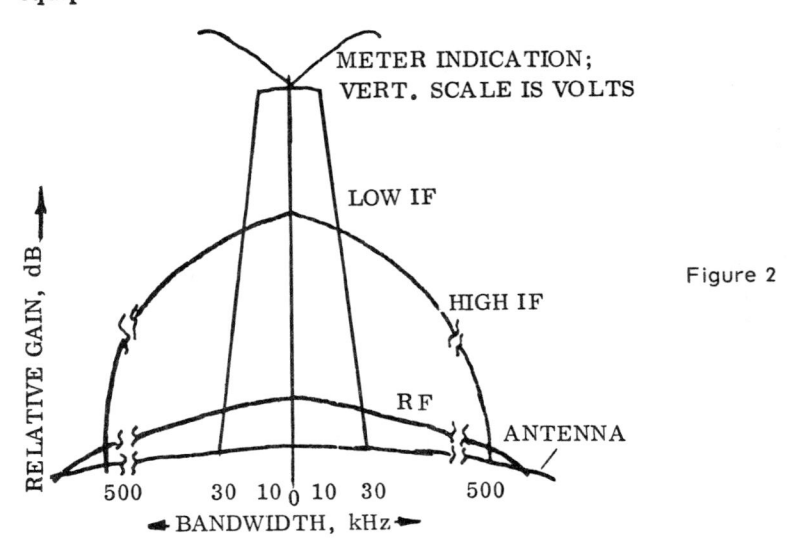

Figure 2

commercial service, the most common source of i-f misalignment is unnecessary tweaking on the part of an incompetent or inexperienced serviceman.

Realignment is usually required if the i-f passbands are not centered on the incoming signal of interest, if the passbands are asymmetrical (not the same on both skirts), or if the bandpass is too narrow. The presence of high impulse noise on weak signals is one symptom of an off-frequency passband. This is due to the fact that the ringing frequency of the filters is not coinciding with the discriminator center frequency. An even more apparent indication of this type of misalignment is "chopping out" of signals or undue distortion of signals which are being deviated at a near-maximum level. The chopping-out effect is the sudden vanishing of a strong signal with each voice peak.

Off-frequency filters also usually produce a high discriminator "idle" reading. If an inexperienced tweaker has been at work, though, he has probably already compensated for this condition by changing the discriminator to get a zero indication—and thereby throwing the receiver even further out of alignment.

So, what do you do when you're certain your receiver needs alignment? The first thing is to be **doubly** sure. If you've no doubts, then get a signal generator and start warming up the receiver. If your receiver is equipped with AFC, disable it. Set up the signal generator to produce a stable signal on the operating frequency, and keep it well below the limiter saturation point.

For units which use double-coil i-f transformers (such as GE and DuMont), the resistor loading method is perhaps the most effective means for obtaining a good receiver alignment. This procedure is a bit complicated but not too difficult. Remember to keep the input signal dead on frequency and below saturation. Tune each stage to the exact peak as described below, then repeat the entire sequence.

The response of an i-f transformer changes with the loaded Q of its resonant circuits. By loading one of the coils with a resistor, its response is lowered to a nonresonant point. The undercoupled coil can then be tuned for maximum deflection of a meter on the first limiter. If the coil is coupled to other coils immediately adjacent to it, both adjacent coils must be similarly loaded.

The value of the resistor must be low enough to produce a sharp peak during tuning, but not so low as to make precise tuning difficult. (The lower the resistance, the broader the peak.) Keep the resistor leads short enough to prevent the introduction of stray capacitance into the circuit; and peak one coil at a time.

There are other methods for alignment, but the above procedure is probably the most satisfactory for the amateur, where the preponderance of such test equipment items as oscilloscopes and sawtooth generators are the exception rather than the rule.

If the discriminator needs adjustment and you are set up with a crystal-controlled i-f generator of some kind, you're in business again. (The generator must be capable of holding a test signal to within 100 Hz of the low i-f.) The procedure described here is not applicable to all discriminator circuits, but is ideal for receivers using Foster-Seeley discriminators (GE and DuMont again).

First, monitor the discriminator current with the proper test meter (0-50 uA for DuMont and 0-2.5V for GE). If possible, disable the second oscillator to prevent receiver "garbage" from causing erroneous readings. Apply a low i-f signal to the first-limiter input and adjust the signal level to saturate the second limiter. Then tune the secondary of the discriminator transformer for a near-zero meter reading, and repeak the primary.

Move the test signal 10 kHz above the low i-f and note the reading; then move it the exact same amount below the i-f. If the readings don't deflect the meter the same each side of zero, adjust the primary until equalization occurs. You'll have to rezero the secondary and adjust the primary several times to make certain the discriminator is properly aligned.

The author gratefully acknowledges the assistance of Donald A. Milbury (W6YAN) and the General Electric Company in preparing this material.

Checking Crystal Ovens

by Jim Lev K6DGX

Frequency stability is of the utmost importance to the serious FM operator. This article is intended to help you achieve the highest stability your equipment is capable of—commensurate, of course, with the grade of crystal you use.

It is a waste of money to buy top-grade commercial-standard crystals only to put them into an oven of unknown or doubtful condition. Why risk damage to an expensive crystal or tolerate transmitter or receiver drift up, down, and around because of a faulty oven? I am employed by a Los Angeles area two-way communications service center where I build and service repeaters as well as mobile telephone equipment. With two or three thousand bases and mobile units in service (quite a few of these are on 450 MHz operating through narrow-band repeaters), we certainly cannot afford to take chances with junkbox ovens!

The method we use to check out crystal ovens is quite simple and more than adequate. First, get a YSI precision thermistor (approximately $4.95). This device comes complete with a graph of resistance versus temperature and covers a broad range from −80 to −150 degrees Celsius at a tolerance error of plus-or-minus 1 percent. It has a resistance of 3000 ohms at 25 degrees and is small enough to mount **inside** an old F-605 crystal case. (In case you've no local supplier, the device may be ordered from Newark Electronics Corporation; their stock number is 29F203.)

Next, unsolder an old crystal holder and mount the thermistor inside. Be certain that it is freely suspended and does not touch the walls of the case.

After installation, resolder the crystal holder together. The device may now be plugged into an oven that is to be tested.

How To Test

Connect an ohmmeter across the appropriate oven pins. It is best to use a good electronic voltmeter in the x10 range.

Apply power to the unit and allow the oven to warm up and stabilize. (This may require five minutes.) By use of the graph supplied with the crystal, determine the center temperature in degrees Celsius (centigrade) and the overall temperature drift caused by the oven cycling. Compare this center temperature with the value expected of your oven; most ovens are 85-degree types, but you may run into an occasional 75-degree type or other "oddball" unit.

Normally, the center temperature of a good oven will be within 2 or 3 degrees of the value specified, and the overall drift will be less than 3 degrees.

So check those ovens. But from now on, be wary of friends bearing gifts: There should be a sudden surge in the availability of bogus ovens!

Ni–Cads—How Not to Ruin Them

There is a great deal of misunderstanding about the charging requirements of nickel-cadmium (or ni-cad) batteries. Manufacturers of such power packs also supply special chargers that "must" be used "exclusively" to avoid damage to the cell.

Many manufacturers of tiny transceivers and pocket communications receivers supply ni-cad batteries with their equipment that likewise "must" be used if their guarantees are to remain valid.

Ni-cad batteries are supposed to be the toughest and best available; they're by far the most expensive. So what is it about them that requires such delicate attention? Why can't any charger be used for ni-cad power sources? The truth of the matter is that two important rules governing use of a ni-cad battery must be observed at all times: (1) Don't overdischarge and (2) don't overcharge.

Overdischarging

Overdischarging can be defined as discharging a cell to the point where it cannot again be fully energized. It isn't easy to know when a nickel-cadmium battery needs charging, because it **should** be charged while there is no noticeable drop in output energy. A battery's output capability (or energy storage capability) is called its "depth of charge." A 100 percent depth of charge is a term applied to a fully charged—usually new—battery. Once a battery has been discharged, it will never again regain the 100 percent depth of charge of its original state, although the level may be imperceptibly below that point. With a nickel-cadmium battery, the further the depth of charge is eaten away during use, the lower the final depth of charge after reenergization. A good rule of thumb is to never allow a nickel-cadmium battery to be used past the 40 percent depth-of-charge point. An even better rule is to keep the depth of charge above 80 percent at all times. In terms of voltage, the point at which a ni-cad battery should be charged is the point at which the per-cell voltage drops to 1.1V under load. (The fully charged cell will measure 1.25V under the load for which it was designed.)

How do you determine the cell voltage of a particular battery? Simple. Measure the minimum-load voltage of the battery when it is fully charged, and divide that number by 1.25; the result will equal the number of cells in the battery. Thus, a 6V ni-cad will measure 6.25V at full charge and will contain five individual cells.

If the depth of charge is kept above 40 percent, a ni-cad cell is easily brought back up to the "95 plus" percentage point time after time (although each charge is slightly—almost immeasurably—less than the preceding one). And if the depth is maintained above 80 percent the charge-discharge cycles can be repeated numberless times while the battery remains in a like-new condition. If the battery can be considered "discharged" at 80 percent, the fluctuation of potential on the battery is minimized, remaining pretty much the same whether the battery is being charged or discharged.

On the other hand, if the energy from a nickel-cadmium battery is completely expended, it stands a good chance of being permanently damaged. At best, overdischarge can prevent the cells from being recharged fully. The battery's efficiency—even at a 70 percent depth of charge—will diminish to the point where the charge is lost at an increasingly rapid rate. A vented ni-cad battery (often improperly called a wet cell) offers a little better rejuvenation potential at full discharge than a nonvented (sealed) battery because of the suspension of the electrodes and the inherent capability of the vented battery to eliminate gases and accept new electrolyte. But even the vented battery is heavily penalized by overdischarging. A completely depleted cell may be brought back up to an 80-85 percent depth of charge again and again after a single overdischarge, but that important top 15 percent may never again be attained.

Overcharging

The most important parameter of a ni-cad battery is its milliampere-hour rating. It is almost impossible to provide a proper charge without having at least a fair idea of what the rating is.

The milliampere-hour rating does not stipulate how many milliamperes of current the battery will provide for one hour. Nor does it tell how many hours the battery will last at a 1 mA drain. The rating is based on this ratio, but the actual figure is calculated to show overall energy output capability to a specified end-point (usually 1.1V per cell) over a 10-hour period. The 10-hour figure is used because the battery's capacity depends on rate of discharge. Because of heating and internal losses, a 100 mA-hour battery wouldn't have the capability of producing 100 mA for a full hour. Yet, it would be likely to produce even more than 1 mA for 100 hours. Thus, the 10-hour standard has been accepted by the battery industry as an inflexible value.

Time is also an important factor in determining length of charge to attain proper energy storage. For practical purposes, the longer the charging time (or the lower the charging current), the higher the resultant depth of charge. Of course, this is only true to a point because there is a practical limit on the depth of charge which can be attained in any case. Generally, an ideal charging time will be more than 10 hours and less than 20.

One might logically deduce that a 250-milliampere-hour battery can be charged with a constant current of 25 mA for 10 hours, 2.5 mA for 100 hours, or 250 mA for 1 hour. The high current rate of the one-hour charge would be as bad on the battery as the high-current discharge rate. Such a high charge rate would almost certainly cause gassing that would wipe out the battery. Another rule of thumb can be applied here: Don't allow the charging current to exceed 10 percent of the battery's milliampere-hour rating, but extend the time period by 50 percent. Instead of charging a 250-milliampere-hour battery at 25 mA for 10 hours, allow it to charge at the 25 mA rate for 15 hours. This will assure that the expended energy is replaced and will allow for various losses and other anomalies.

The lead-acid battery that starts the car each morning is a tough old brute that can be mistreated and manhandled. But even this old workhorse is cheated of longevity when it is given a one-hour charge at its full ampere-hour rating. The ni-cad suffers a great deal more from abuse by overcharging than the lead-acid type. Two or three severe overcharges will destroy a battery or cell that might otherwise have lasted for thousands of charge-discharge cycles.

One other thing to remember: A battery is only as capable as its weakest cell. You may be able to get by with damaging only one cell of a 12.5V battery during a heavy charge. But the battery is just as useless as if they'd all been destroyed.

Gassing

Nickel-cadmium batteries generate gases during the last few hours of charging and throughout most of the cycle during overcharge. Hydrogen forms at the cadmium electrode and oxygen forms at the nickel electrode. Vented cells have removable ports to allow these gases to be freed along with the electrolyte fumes during the charge cycle. But in the sealed ni-cad—the type used in most miniature electronic applications—the gases must be accommodated or used in some way to avoid destruction by overpressure.

Burgess ni-cad batteries are designed so the cadmium electrode has an excess ampere-hour capacity. This feature causes the positive nickel electrode to become fully charged first so it will begin to generate oxygen. The oxygen travels to the surface of the negative cadmium electrode where it reacts to form cadmium oxide. The overall effect is to keep the cadmium electrode oxidized at a rate just sufficient to offset

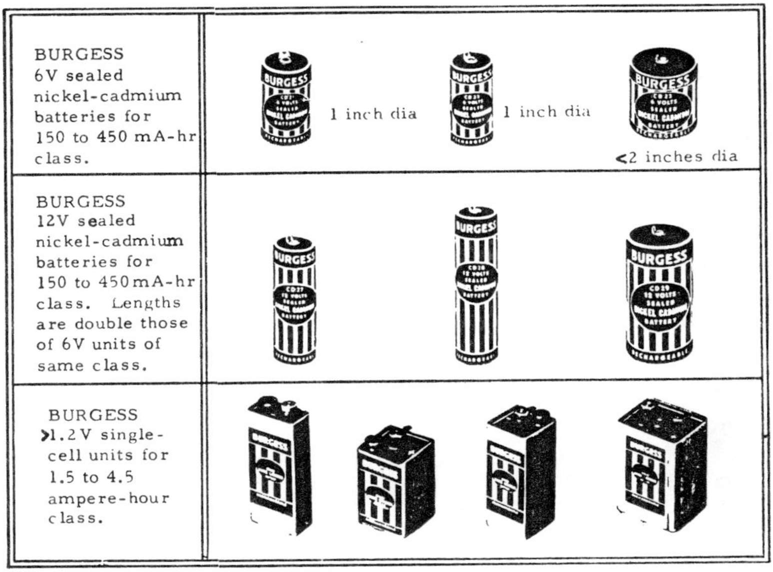

BURGESS 6V sealed nickel-cadmium batteries for 150 to 450 mA-hr class.	1 inch dia	1 inch dia	<2 inches dia
BURGESS 12V sealed nickel-cadmium batteries for 150 to 450 mA-hr class. Lengths are double those of 6V units of same class.			
BURGESS >1.2V single-cell units for 1.5 to 4.5 ampere-hour class.			

Figure 1

the input energy, and the cell is maintained at a reasonably stable equilibrium at full charge.

But even with this precautionary measure, overcharging can damage the cell. High-rate charging can cause oxygen to be produced at a faster rate than it can be used at the cadmium electrode. This can cause pressure buildup to the point where the seal is ruptured.

Charging

Knowing the milliampere-hour rating of a ni-cad battery is extremely important if its life is to be protected. If there are no clues provided on the battery case, the rating can usually be determined within a fair margin of error by estimating. A standard D-size 6V cell (the size of a conventional flashlight battery) will have a milliampere-hour rating of approximately 250. Using the 10 percent rule, it can be seen that the basic charge rate is 25 mA; and by application of the "plus 50 percent" time rule, the proper charge period is 15 hours.

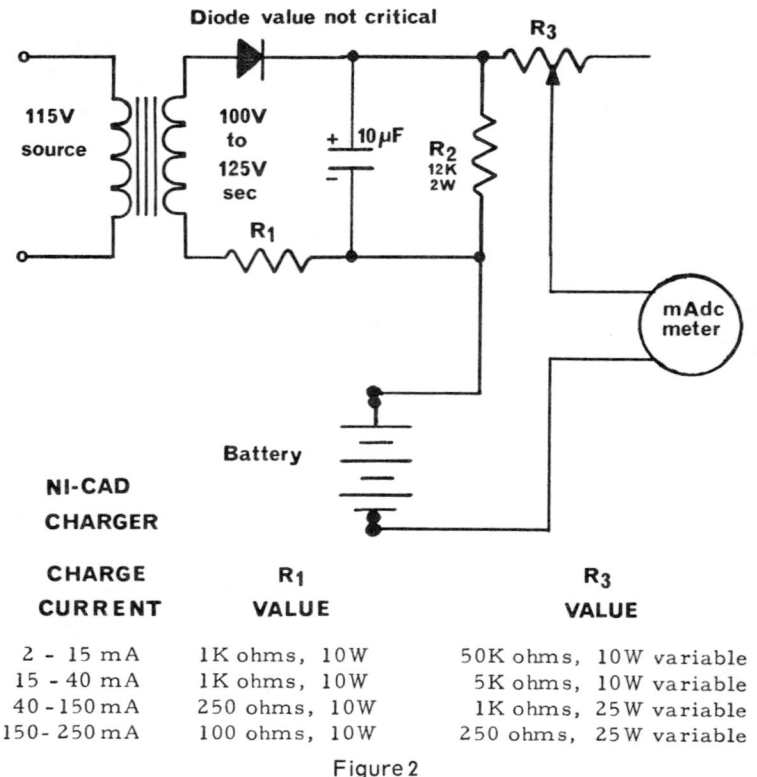

NI-CAD CHARGER

CHARGE CURRENT	R1 VALUE	R3 VALUE
2 - 15 mA	1K ohms, 10W	50K ohms, 10W variable
15 - 40 mA	1K ohms, 10W	5K ohms, 10W variable
40 - 150 mA	250 ohms, 10W	1K ohms, 25W variable
150 - 250 mA	100 ohms, 10W	250 ohms, 25W variable

Figure 2

Trickle-charging may be employed if the battery is used at low drain rates. A general rule for trickle-charging is to maintain the charge level at 10 percent of the standard charge rate, and keep the battery under this charge during all periods of nonuse. The trickle-charge current for the D-size battery is 2.5 mA. A very simple battery charger can be built up readily with run-of-the-junkbox parts. If the battery is not to be in use during a charge, a half-wave rectifier will be adequate. The diagram on the preceding page shows a simple rectifier circuit and lists the component values for various charging currents.

How to Get the Most from Your Mobile

by Bill Harris K9FOV

There are two pertinent factors upon which the successful operation of a land-mobile radio station hinges; they are: (1) quality of installation and maintenance, and (2) operator technique. This article will endeavor to help you with the first.

You probably already have at least one mobile unit in the family ornithopter, but perhaps you're contemplating the addition of another. Or possibly a change of cars is in the offing. Perhaps you're a new FM'er with a fresh unit and you're preparing to install your first unit. In any case, some of the hints here might be good to keep in mind at installation or maintenance time.

The first item of note has little to do with actual installation, but it will have a great deal of bearing on the operation and performance of your unit; that item is the crystals you intend to use for transmitting and receiving. DON'T TRY TO SAVE A BUCK OR TWO ON CRYSTALS! What you may have saved in dollars will not buy back what you will have lost in performance through the use of substandard crystals. Twelve or fourteen dollars for a set of stable crystals may seem high, but bear in mind that it will be a one-time investment; they'll last the lifetime of the rig, and the frequency will be the same the day your radio dies of old age as it was on the day of its birth. The rule is: Get the crystals made for the unit and you will never have frequency stability and drift and "badrock" problems that might otherwise cripple the rig. And NEVER try to "fake" the crystal with one you just happen to have lying around the shop. With conventional ham radio, such tactics are clever; with FM, they're crude. Quite understandably, it's the most common mistake FM newcomers make. If you're a newcomer and you've committed this sin, rectify it now by ordering your trouble-free crystal and slipping it quietly into place when it arrives. Do it without fanfare and no one will ever know.

It goes without saying that the mobile rig should be checked completely before installation, and all noticeable defects corrected at that time. In this way, you can spot such problems as broken cables, loose connections, and the like while you can still get at them. This can prove to be a pretty handy tip if there's no one within 80 miles of you who has the necessary jig and cables to fire up your unit on the bench. Troubleshooting in the trunk of a car tends to be a time- and gas-consuming headache, especially when the difficulty is serious and the light is poor.

You can help to overcome the potential problem of inadequate lighting by taking the time to install a simple light on the inside of the deck lid. Most auto parts stores sell mercury-switch lights that can be installed in minutes; such lights go on when the lid goes up and go off when the lid is shut. (At least the instructions say the light goes out when the lid goes down; it would be an interesting object of research to check the validity of the statement. Maybe I'll do some study along these lines, but it will have to wait until I'm finished with my current "refrigerator-light" effort.)

Look at your car as a rolling QTH—not necessarily as an "unspoiled thing of beauty and a joy forever" when installing the radio. Unless you use it only to drive back and forth to church once a week, it's going to show the inevitable signs of wear soon enough anyway. So at least give some consideration to mounting the antenna in the roof if at all possible.

Contrary to popular opinion, the small hole in the roof does not depreciate the value of the car these days. And if you don't believe that statement, ask any car dealer. If it's the effects of the hole that bother you, take heart: A good make of antenna properly installed will not leak. Unless you use a car-top carrier a good part of the time, the space up there is going to waste. So why not use it—you won't regret it.

Even if you're a six-meter FM'er, a roof mounting antenna will be less directional than a rear-mounted ball and spring.

Never use a bumper mount on six meters. Inherent band noise is worse on six than it is on any other amateur band. A bumper-mounted whip compounds the problem in two ways: it puts the antenna close to the ground (which is a noise source in itself) where signals find it difficult to compete with road garbage; and it hides the antenna behind the mass of the car. On any other rear-mounted type of installation, the car mass reinforces the signal by lending it gain and directivity off the front (at the sacrifice of side coverage, of course): but this is not so with the bumper mount.

Other types of antenna installations to stay away from if you're making it a permanent hookup are gutter clamps, jiffy mounts, trunk-groove and magnet-base mounts and their ilk. For the most part, they give misleading vswr indications and result in high radiation angles due to their self-resonance and inherent inductance in the ground return. Also, they tend to leave the transmission line and coaxial termination out in the open where eventual deterioration is inevitable. To top it all off, the kink in the cable where the door, window or trunk lid closes will ruin it in short order.

And while we are discussing the cable aspect: Upon installation and periodically thereafter, carefully check the transmission line; replace it if it shows signs of degradation, crushing, or right-angle bends. It may not show up on a vswr meter, but weather contamination or center-conductor migration will wipe out a lot of performance.

When installing the unit, fasten it firmly to the floor. In the case of a front mount, get some heavy trunnion brackets made and use them to bolt the case to the dashboard (and to the floor or firewall, if possible). There are several sound reasons for mounting the rig securely: For one, the same amount of rf current that goes to the antenna must flow through the car body, and it does not generally prefer the dynamotor ground lead or the coax braid for its path to ground. The battery current WILL, however, tend to return through the coax braid for its ground path, particularly if the negative cable connection happens to loosen a bit. This will tend to affect the receiver sensitivity due to increased vibrator hash pickup along the lead.

The last reason I wish to expound in favor of fastening the rig securely is one that has probably never entered your mind: Say you are involved in a wreck and you hit something head-on or roll the car a time or two. What will you have accomplished if you are kept intact by the seat belts only to be decapitated by your faithful old 80D (which weighs something like 1400 pounds at 60 miles per)?

Inside the cockpit, mount the speaker where you can look directly into it when you're sitting in the driver's seat. It's usually better not to give the preferential location to the control head, and stash the speaker way up under the dash panel where it can talk to all the defroster ducts and wiper cables. Since the control head is a set-and-forget device, it should be the item mounted in the more remote spot. In any case, intelligibility of signals will be greatly improved by judicious situation of the speaker; and likewise, intelligibility will be seriously degraded if the speaker is not mounted reasonably close to the listener, and positioned toward the listener's ear.

If the speaker has a warped or torn cone, replace it. It's an inexpensive item to buy, but it is still one of the weakest links in your radio system. In addition to affecting the readability of signals, a speaker with a rubbing voice coil will actually increase the apparent noise in the audio. (Standing waves here, too.)

The mike cord deserves some consideration, too. What could be more hazardous than snagging the mike cord in the wheel as you tool around a tight corner? A little thought goes a long way in terms of safety, comfort, and convenience.

Sit in the driver's seat and close your eyes; angle your right arm straight out and lay your palm on the dash. That's the spot where the mike bracket should be mounted. Unfortunately, in 98 percent of the newer cars, this spot is infeasible, so try to find one that is not. In any event locate it where the operator can grasp it without taking his eyes off the road.

Here's a helpful hint on cables: Remove the cables from the battery and carefully clean the connectors and the battery posts with soapy steel wool and warm water. At the same time, it's a good idea to clean off all foreign matter from the top of the battery. Rinse the area well, then dry it off and reconnect the cables to the battery. When the reconnection is completed, spread a thin coat of silicone grease over the connection; chances are you'll never have to wade through the oxidation again (and neither will the current to the battery from the generator or from the battery to the radio or starter, I might add).

Don't forget an ignition tuneup or any other noise suppression that may be necessary. (There are many articles and books available on this subject.) Of course, FM is not as susceptible to noise as some of the other modes, but it is by no means immune; it's just that you don't notice noise presence so much because of the squelched receiver.

After the unit is installed and peaked to the antenna, check with someone to make sure you are transmitting on frequency; if warranted, make any necessary adjustments. The closer you get to frequency, the stronger your transmitted signal will seem. At this time, zero the receiver oscillator onto channel by monitoring a known signal with a discriminator meter. Get a few modulation reports, and adjust the transmitter deviation as necessary.

This should find you all set to go FM mobile. Just remember, you'll only get out of your unit what you have put into it!

FM
Service Center
by Don Chase W0DKU

In both Motorola and GE high-band transmitters using triode modulator stages, if you have low power output, but all voltages and drive levels check out okay, look for high power factor in the electrolytic capacitor from cathode to chassis ground in the modulator stage. If the power factor is high, even though you have adequate grid drive to the final, the power output will be low.

*

If you come into possession of a very dirty radio, take it to the local "do-it-yourself" carwash. Use the hot detergent generously; just be sure to rinse very thoroughly with clear water. Then take it home, set your oven on its lowest heat (usually 150-175 degrees) and bake for four hours. Any parts that would be damaged by this treatment would need replacement anyway. Just be sure the equipment is completely dry before applying any power. GE technicians even recommend washing transmitter and receiver strips in the family automatic dishwasher.

*

Tight-fitting slugs in your equipment? An easy way to lubricate them is to use one of the silicone spray products for lubricating sticky drawers, available at most hardware stores and supermarkets.

*

GE's "Message Mate" is made only in a narrowband version. By carefully stagger-tuning the low i-f filter (the three cans immediately following the second mixer), the units perform quite well on wideband.

*

In severe cases of ignition noise in your mobile unit, check the polarity of the ignition coil. Cases have been found that required nothing more than swapping the primary leads to bring the noise down to a listenable level.

*

Using the 6146B in 12-volt mobile transmitters? Check the filament balance: The 6146B draws less current than the 6146 or 6146A.

*

Are you having trouble with spurious emissions from the transmitter of a GE Progress Line unit "locking up" your GE repeater? Two sources of trouble are the 6678/6U8A tube (modulator) and the 6677/6CL7 tubes. The 6678/6U8 can be replaced by a 6GH8, which usually seems to cure this problem.

*

Here's a hint if you're having trouble neutralizing a high-power two-meter power amplifier: Try hooking the output of the driver to the output of the power amplifier; then remove the B+ from the power amplifier and watch the grid current of the power amplifier as you adjust the neutralizing capacitor.

Defeating Desensitization In Repeaters

by Van Fields W2OQI

In a typical repeater situation, the transmitted signal causes the limiter current on the receiver to increase (thus affecting receiver sensitivity at the time when sensitivity is needed most). It appears that the rf and mixer stages are biased class C and the mixer must generate noise that the limiter "sees."

In the diagram, the transmitted signal is transmitted from the lower antenna and introduced into the upper receiving antenna. The idea behind this scheme is to introduce the same signal back 180º out of phase and at the same amplitude. This is achieved by sampling the transmitted signal with a directional coupler, adjusting the phase with the line stretcher, and adjusting the critical length to the cavity.

The cavity has a small loop that is variable so the attenuation may be adjusted with no phase shift. Ordinary attenuators also introduce varying amounts of phase shift, thus giving two variables to be adjusted at once.

Motorola normally puts in three large cavities with small loops to get the high skirt selectivity (and high losses). They usually shun spacings this close in frequency.

To adjust, watch first-limiter current on a sensitive meter and turn adjustable cavity loop to about 45º. Next, adjust the line stretcher. A dip should be noted on the meter at some point. If the dip comes at the end of the adjustment (and it always seems to), add a small section of coaxial cable.

It pays to have several short random lengths of coax available (or you can cut each one a nanosecond longer than the last).

Once you have a dip, adjust the cavity loop for minimum limiter current.

Once it is operating, the line stretcher can be replaced with a piece of coax of the proper length. Trimming can be accomplished by adding connectors or adapters.

It is necessary to keep the two antennas rigid so they won't move with the breezes characteristic of the locale. But more importantly, the system must be moisture-proof. If the vswr shifts on a damp night, the phasing is out and so is the repeater. Ideally, one antenna with a ferrite isolator would be used, but as yet no isolator has been obtained for 146 MHz.

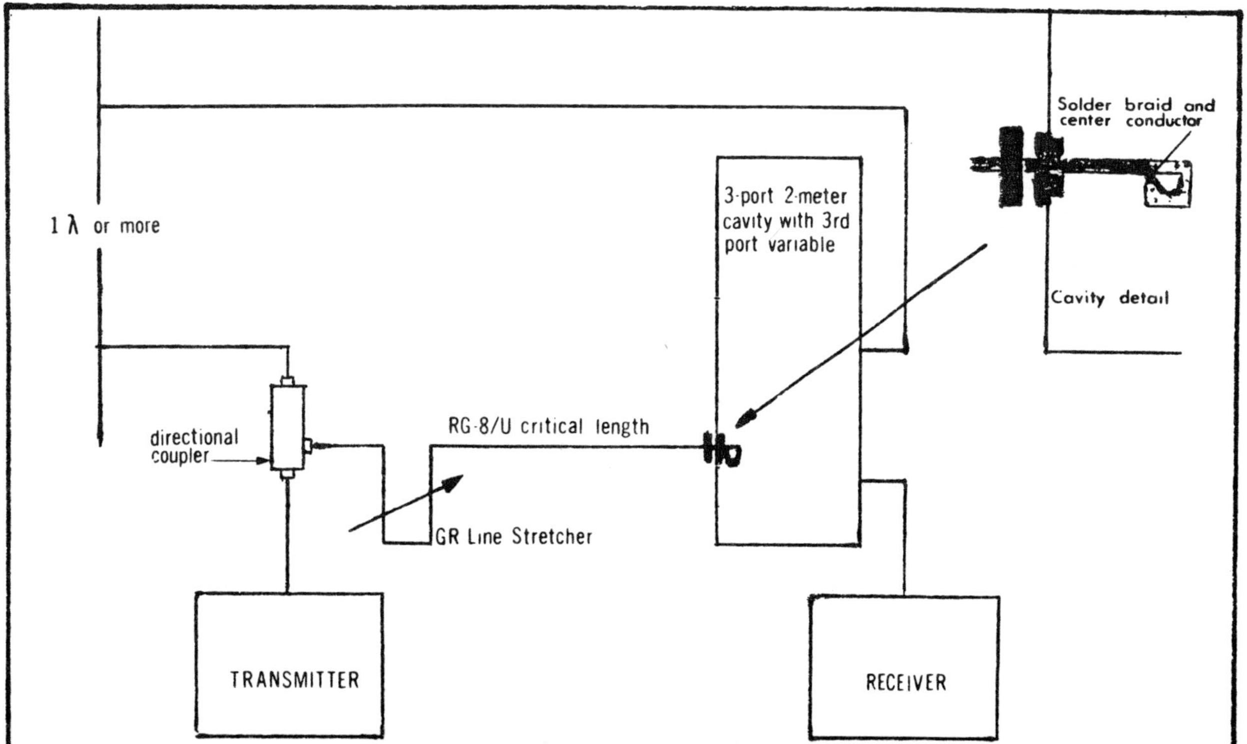

Figure 1

Motorola Permakays

Thanks to the California Amateur Relay Council for providing this complete list of Permakays for FM readers:

	MOD ACCEPT	CHANNEL SPACING	RECEIVER	FREQUENCY BAND
F208	5			
K8435	7½	20/30	A	1/3
K8436A	15	40/60	A	1/3
K9035A	15		C	
K9076		AM Recvr	PA9077	1500-3000 kH
K9135	30	120	Unichannel	3
K9240	5	20	G	1
K9241	15	40	D G	1
K9242	30	120	G	1
K9245	5		A	
K9341	15		A	
NFN 6000AW	15	40/60	Dispatcher	1/3
NFN 6000AS	5	20/30	Dispatcher	1/3
NFN 6001AW	15	40/60	VHF Pager	1/3
NFN 6001AS	5	20/30	VHF Pager	1/3
NFN 6003AW	15	40/60	NRD 6112 Series	1/3 (H21-23 DCN)
NFN 6003AS	5	20/30	NRD 6112 Series	1/3 (H21-23 DCN)
NFN 6004AW	15	40/60	NRD 1130 Series	1/3 (Portable)
NFN 6004AS	5	20/30	NRD 1130 Series	1/3 (Portable)
NFN 6005AW	15	50	NRE 6000	4 (H24DCN)
NFN 6006AW	15	40/60	NRD 6111 NRD 6112	1/3 (H21-23 DEN)
NFN 6006AS	5	20/30	NRD 6111 NRD 6112	1/3 (H21-23 DEN)
TFN 6000AW	15	50	U44BBT	4
TFN 6000AS	5	25	UHF B	4
TFN 6001A	15	50	UHF B	4
TFN 6004AX	30	120	Unichannel	3
TFN 6007AS	5	20	A	2
TFN 6008AW	15	40/60	H - C & D Motrac	1/3
TFN 6008AS	5	20/30	H - C & D Motrac	1/3
TFN 6013AW	15	40/60	H - C & D Motrac	1/3/4
TFN 6013AS	5	20/30	H - C & D Motrac	1/3/4
TFN 6014AS	5	30	T1230AH Cartelephone	3
TFN 6015AS	5	20/30	C Business Dispatcher	1/3
TFN 6017AW	15	40/60	H - E Motrac L & M Motrac/Motran	1/3
TFN 6017AS	5	20/30	H - E Motrac L & M Motrac/Motran	1/3
TFN 6017CS	5	20/30	M Motrac—Motran	1/3
TFN 6018AS	5	30	IMTS Cartelephone	3
TU 145	15	60	A D G	A-4, D&G-3
TU 145A	15	50	A G	3/4
TU 194	15	50	A	4
TU 322	15		A	4
TU 344	40	100	A	5
TU 406	5	20/30	A	1/3
TU 455	15	30	G	3
TU 456	30	120	G	3
TU 540W	15	40/60	G	1/3
TU 540S	5	20/30	G	1/3
TU 540X	30	120	G	1/3
1V847945	15	40/30	Early Dispatcher	1/3

Deviation Setting by Clever Estimating

by Bill Harris K9FOV

An overmodulated FM signal may sound raspy to one listener and clear to another. The human ear tends to be a rather inconsistent indicator. Setting transmitter deviation by reception reports, though sometimes necessary, is seldom a satisfactory approach.

If one uses a good FM receiver to monitor for distortion, and employs the ear merely as a "go/no-go" indicating device, the adjustment can be made with at least a "passing" degree of accuracy.

Take a properly operating receiver such as Motorola, GE, or RCA that is **known to be in good alignment** and sensitive. Good alignment is an ABSOLUTE NECESSITY! The receiver should be one on which the squelch will open fully on a signal that is not quite full quieting. Key the transmitter in question within the local coverage area of the receiver so as to provide a full-dead-quieting signal in the receiver. Check to be sure the transmitter is netted to the receiver to within 1 or 2 microamperes. Whistle about a 1 or 1.5 kHz tone loudly, cleanly, and continuously into the mike, while adjusting the deviation control until the squelch cannot be quite "whistled closed" in this manner. This method even compensates for the presence or absence of audio rolloff filtering in the transmitter audio, since the receiver noise amplifier actually measures overall distortion of the received signal. Thus, the transmitted deviation will closely approximate the measuring receiver's bandwidth characteristics.

I have bench-checked many rigs, both wide and narrow, that have been set in this manner, and can report better than 90 percent accuracy, with all errors being on the low side rather than the high side.

GENERAL TECHNICAL INFORMATION

The Case for Narrowband

by Robert Kelty W6DJT

Recent discussions concerning operation of amateur relay systems at the 5 kHz narrowband standard have evoked not only considerable constructive debate but some misinformation as well. Within two years or less, surplus FM mobile equipment availability will be the determining factor as to what deviation amateurs use in FM systems just as surpluses of 450 MHz equipment available will encourage continuing expansion into UHF.

But what is wrong with what we have operating now? Why change anything? Everything is working okay now, so what's the difference? All quite true, and I would be last to suggest wholesale conversions for no reason. But let's examine the situation facing us so that we have a better appreciation of what is coming in amateur FM systems.

Historically consider the progression of events in FM mobile since we are directly affected by using equipment built to standards for these services. Up until 1950, deviation was not controlled very rigidly. Channels were in 120 kHz spacings, deviation limiting was not common, stability was not important (pressure-mounted crystals were the rule), receivers were wide as all outdoors—even purposely swamped by resistors across i-f circuits, and no attempt was made to control audio frequencies transmitted. Listening to these old systems was almost like listening to a broadcast receiver; and the manufacturers must have thought similarly since large soft-cone speakers were a part of mobile installations.

But radio became popular in more than police cars and commercial vehicles. The spectrum had to accept users from areas that previously had never considered radio. And regulations tightened. In 1950, deviation limiting was required of all older equipment—at the figure of 15 kHz to permit users to occupy channels only 60 kHz spaced. It was quite apparent that equipment was to be asked to do more. Manufacturers looked ahead and saw what was coming with an ever-increasing number of spectrum users.

In 1952, the first equipment was produced that could be readily converted to a tighter expected standard, an inevitability dictated only by the economics of a nationwide program. Narrowband was the scheme; it meant a four-year period of preparation, replacement of all older gear which could not be economically converted, and design of more modern equipment to meet requirements. But the impetus given by a plan of this nature meant more than just spending dollars so that more users could occupy the radio spectrum. It encouraged better equipment development, improvement of techniques, and considerably upgraded systems. It spelled the end of Loktal tubes, pressure-mounted nonheated crystals, and when the 12-volt V8 cars became the standard in 1955, it meant an entirely new approach to FM mobile. The wideband system was to be as out of date as the flathead 6-volt V8 cars or AM radio itself.

In 1956, the plan was implemented and conversions began. Large quantities of older, nonconvertible gear were removed from service—equipment that may have had more years of life. But when 12-volt conversions were added to narrowband conversions, economics ruled and replacement was the only answer.

Up to this point, the amateur concerned himself more with AM or sideband, while FM was about as uncommon as amateur TV. But we are not well known for letting surpluses go to waste nor to disregarding an economically feasible idea. True to style, an imaginative group procured a few mobile units, and although the idea of going FM was not new, they built an FM mobile relay, located it at an advantageous location, and started a whole new field of endeavor.

Today the amateur FM relay has gone through several stages, and equipment available to affluent and knowledgeable FM enthusiasts is certainly quite different from the way it was 10 years ago. Rather than Loktal-tubed, pressure-mounted-crystal units being the grade of surplus available, we find narrowband, heated-crystal, transistor-powered equipment the thing.

Converting equipment back to wideband operation for amateur use has become a matter of concern and an economic and time-consuming problem yet to be resolved. Differences in wide and narrow systems were discussed by land mobile people in the early 1950s, material was widely published during conversions in 1956-1958, and resurrected by concerned amateurs facing conversion of systems in 1967-1968 when narrow standard equipment became readily available. But the facts have not changed one iota. Let's look at them on a practical basis.

FCC Amateur Service Regulations are sketchy at best for FM deviation. Wideband is not specified, narrowband is described as "occupying a bandwidth not more than a comparable AM signal of similar audio characteristics." A calculated guess at what this means is ±3.0 kHz, a figure totally impractical for FM mobile equipment built for either the ±5 or ±15 kHz standard. A ±2.5 kHz standard was considered and rejected in 1958 as being out of the question for land mobile use. On the other hand, wideband deviation could be ±20 or even ±30 kHz and not be in violation.

How does equipment differ in wide and narrow systems? Transmitters of modern design have deviation limiters and postlimiter filters which roll off audio at a predetermined rate. Deviation can be kept at maximum permissible level, 5 or 15 kHz, without overshoot or failure to occupy the full channel width. In narrowband systems, transmitter deviation is not merely turned down. The postlimiter filter for audio rolloff also is a simple but important part to tailor audio and prevent splattering the adjacent channels with higher frequency audio. In one's own system this splatter could be seen in receivers as high frequency noise; it serves to close the squelch and chop audio. All transmitters built since the 1952-1954 era usually have these desirable characteristics. Receiver differences are equally simple and are comprised of circuit changes for audio recovery, bandpass acceptance, and in some cases squelch circuitry. Discriminators in narrowband equipment must recover 1/3 the audio of wideband systems and are commonly converted by tripling the value of two diode-load resistors or otherwise increasing audio passed onto audio stages. Bandpass acceptance of receivers is modified by replacement of a passive filter in some equipment while in others the change is made by altering the 2nd i-f transformer coil coupling. In some cases, a minor change in squelch circuitry changes biasing to balance relation of a fixed dc voltage and rectified high frequency noise which serves to close the squelch. In all cases, receiver sensitivity increases in narrowband operation, the squelch takes on a more positive operation, and that indefinable sharper sound is noticed. Although some older receivers cannot be converted down to a very sharp 5 kHz modulation acceptance, such gear as Link 1905-1907 (when second i-f swamping resistors are removed) are significantly sharper and narrower, tune with sharp peaks, and improve in sensitivity. Others approach the desired bandpass for acceptable performances. All transmitters and receivers in narrowband systems have heated crystals or otherwise stabilized frequency-determining elements if equipment is that modern. And, of course, the pressure-mounted crystal is out of the question due to tolerances and drift. One of the more important requirements in modern systems is maintaining all transmitters and receivers on the same frequency—easily done with crystal ovens and quality crystals. Transmitters and receivers should be within about 750 Hz of designated frequency but designs are such that when both drift in opposite directions (the worst possible case), no severe misoperation occurs.

Reviewing all differences and requirements detailed for transmitters and receivers might appear to be a monumental task for the individual but when it is considered that all current properly operating equipment already has these desirable characteristics built in, the situation is not so severe. The conversion kits for nonconverted equipment are simple and inexpensive. Whether using the 5 or 15 kHz standard is a matter of preference, as long as the manufacturer's circuitry is used, that great gray area of figuring it out for yourself is eliminated.

What about equipment performance? First of all, receiver sensitivity increases somewhat. As noted earlier, squelch operation is less sloppy while sharper performance is noted. Possibility of interference from off frequency or AM stations is reduced by virtue of occupancy of less spectrum and the fellow who tries to whistlestop-tune on channel has a harder time. In the case of older receivers which were never intended to operate on narrow standards, conversions noted above are readily accomplished if only partially.

Unless receivers are converted for better audio recovery (wideband discriminator recovering narrowband audio) there could be marked difference between squelch tail and audio amplitude. Some receivers—GE Progress, Motorola Research, and Link 1905— had audio recovery which was quite adequate for narrow systems. The two-resistor discriminator modification is usually not necessary on these units, but bandpass is another matter. Consider a typical i-f-swamped wide receiver to be operated narrow. Unless the signal is quieting the receiver, an undesirable condition can occur in that only 1/3 of the bandwidth in the wide receiver is occupied. The signal-to-noise ratio can be intolerable; intelligibility is degraded. Relay systems usually give adequate signals to receivers if no adjacent channel users are on the air. Ideally, however, the user should try to occupy all of his receiver bandwidth. In the overly broad receiver conditions described above, ignition noise can be a nuisance, but when bandwidth of receivers is modified, intelligibility is less likely to be degraded.

Transmitters in narrow systems, while occupying less frequency than in wide, also produce less splatter and noise. In so doing, an associated receiver working as a part of a relay system is less likely to be interfered with, adjacent channel users receive less splatter, and less spectrum is occupied per user, really not of significance to the nonconservation-oriented amateur. Despite a tendency to think in terms of "just turning down" deviation, any conversion to narrowband operation should include the other minor requirements mentioned earlier.

And how about performance comparison of wide and narrow systems? The major difference is in signal-to-noise ratio. Narrow bandwidths are more susceptible to ignition noise from high-compression engines without proper noise reduction, particularly at low signal levels. When pushed to the limit, audio recovery in narrowband receivers renders considerably less intelligibility than wideband. In the typical relay system, signals are usually high enough to eliminate this minor difference and performance is usually improved because modern receivers have been designed for narrow characteristics and consequently have better performance. The receiver converted to wideband has been done so at a deliberate sacrifice in sensitivity, but the improved signal-to-noise ratio may more than compensate for this.

Another requirement of narrow systems is better frequency control—commercial quality crystals and ovens. Although Sentry or International crystal cost is generally double that of the average amateur crystal, the equipment into which the high-quality crystal is placed is certainly a grade beyond average amateur VHF gear. Considering the importance of crystals to radio equipment, this is only a small part of costs, remaining with a transmitter or receiver throughout its useful life, not deteriorating such as tubes, and carrying a warranty of performance. How many other parts of a combination is so much trust put into?

Whether choosing a wide or narrow arrangement is immaterial so long as group objectives are met. The choice will depend on equipment available, expected conversion costs, and economics. Despite misinformation available, either bandwidths operate well when all equipment complies with the standard established. No further proof of this is better represented as in the many systems operating in other land mobile services.

If equipment available today is causing you to wonder about what to do, have faith. Most of that older wide equipment can be integrated or converted satisfactorily. A whole lot cheaper, I might add, than widening a MASTR, MOTRAC, TPL, or even Progress (T-power category).

How to Know What to Buy

When you look over the used-equipment catalogs, you will note a marked inconsistency in pricing structure. A late-model DuMont base station, for example, might sell for a fraction of the price of an equivalent Motorola unit. Or a 10-year old GE unit may be listed at many times the price of a near-new ITT.

GE and Motorola command consistently higher prices than other makes. But the higher price does not necessarily mean the GE and Motorola units are superior. GE and Motorola are the two biggest names in two-way radio. But RCA, DuMont, Kaar, Aerotron, and Bendix are common, too. Why, then, are GE and Motorola so sought after?

The amateur buys with the thought of selling, for one thing. And he knows that any amateur with any experience in FM has experience with Motorola—or GE. So when he buys, he tries to buy something other FM'ers have used or serviced.

Another contributing factor is documentation. Handbooks, schematics, troubleshooting procedures may or may not be available from the manufacturer, but every FM'er knows he can get access to most Motorola and GE manuals simply by visiting the nearest commercial two-way service center.

But perhaps the principal reason most amateurs select GE or Motorola is "general familiarity." When an inexperienced amateur runs into some trouble trying to service a GE or Motorola unit, he can feel pretty sure of getting experienced help from other local FM amateurs who've worked on similar units with similar problems in the past. Interchangeability of components might be another consideration in favor of the big two. For a number of years, GE and Motorola (and one or two other manufacturers) have employed the "module" concept in the manufacture of their two-way equipment. For a given model "year," for example, the manufacturer will produce singular transmitter and receiver designs. The transmitter, receiver, and power supply are constructed on individual removable chassis.

Thus, for a given radio model, a transmitter "strip" from a mobile unit may be interchanged with a transmitter from a base station; or a base station receiver "strip" can be used in the equivalent model mobile unit.

The lower price of RCA, DuMont, and the others, however, can make a strong case against the two big names. The ultimate decision, of course, must rest with you, the buyer. If you feel confident that you'll be able to handle your own service problems and that you won't be trying to make a quick sale on your equipment, it might pay to weigh a purchase purely in terms of performance versus dollars. The secondary names in the two-way industry have also manufactured gear of sound design and quality workmanship. (They've had to; look at the competition they've been bucking.)

Identifying Motorola Equipment By Model Number

The model numbers of some manufacturers can provide a great deal of information about the equipment itself. Motorola, for example, has for at least the past decade employed a model numbering system that gives the type of equipment (mobile or base), method of mounting (trunk or dash), rf power, frequency range, vintages of receiver and transmitter, and type of power supply. A familiarity with these numbering systems will help immensely in evaluating a given piece of equipment.

Motorola model numbers assigned after initiation of the descriptive system include six basic characters in this sequence: letter, digit, digit, letter, letter, letter. The first character tells whether the unit is for portable, mobile, or base station use and gives the following general data:

H—Handie-Talkie
P—Hand-held (portable) unit, larger than Handie-Talkie
L—Tabletop base station
D—Dash mounting mobile unit
T—Trunk mounting mobile unit
U—Universal (dash or trunk) mounting mobile unit
J—Tall cabinet, weatherproof enclosure
B—Tall cabinet, indoor use
W—Western Electric (mobile telephone)

The second character (digit) describes the power capability of the unit:

1—1 watt or less
2—2-5 watts (up to 20W input)
3—10-15 watts (up to 50W input)
4—18-30 watts (up to 80W input)
5—30-60 watts (up to 180W input)
6—70-100 watts
7—150 watts output

8—250 watts output
9—300 watts output

The third character (digit) identifies the frequency of operation:
1—Low band (50 MHz)
2—Mid band (70 MHz)
3—High band (150 MHz)
4—UHF (450 MHz)

The next two letters of the sequence identify the model of the transmitter and the receiver, respectively. The last character (letter) identifies the power supply type:
B—Base station (115V ac)
T—Transistor (mobile, 12V)
V—Vibrator (mobile)
D—Dynamotor (mobile)
N—Nickel-cadmium battery pack

Identifying GE Equipment By Model Number

The GE numbering system isn't quite as descriptive as the Motorola procedure. The Progress Line system was an improvement over that of the Pre-Prog, but the almost numberless variations make it impossible to accurately identify a unit by its model number alone. In the interest of completeness, however, Table 1 describes all pre-Progress Line units and shows the differences between models.

Summary

It is difficult to establish rules of thumb regarding the pricing structure of used two-way radio equipment, but there are a few applicable generalizations it might pay to remember: Used-equipment prices tend to be reasonably stable. Motorola and GE units, particularly the later models, seem to stay somewhat in demand, and are thus usually easy to sell. Two-piece units (separate transmitter and receiver) are usually a bad buy at any price; the transmitters often do not incorporate limiter circuits to stabilize the audio output, and the receivers are typically far too broad for effective application where adjacent frequencies are likely to be active. One-piece units (of modular construction) are particularly desirable because they provide the means for "system" tailoring to individual requirements and allow a considerable degree of freedom for component / module interchange. For mobiles, vibrator or transistor power supplies offer a significant advantage (in terms of efficiency and power consumption) over dynamotors.

One additional point, in case you plan to buy one of the "off brand" makes. Be sure the tubes it uses are standard enough

Table 1

Freq.	Number	RF Power Out.	DC Input Voltage	type of mount	type of power supply
			MOBILE		
152-174 MHz	LPH6/12	3W	6/12	single-unit	vibrator
	LPH24	3W	19-30	single-unit	dc-ac conv & ac pwr spply
	LPH32	3W	26-40	single-unit	dc-ac conv & ac pwr spply
	LPH64	3W	60-84	single-unit	dc-ac conv & ac pwr spply
	MC203	10W	6	single-unit	vibrator
	MC203-LP	LPI(1 W)	6	single-unit	vibrator
	MC204	50W	6	single-unit	dynamotor & vibrator
	MC205	35W	6	single-unit	dynamotor & vibrator
	MC206	25-30W	6	single-unit	dynamotor & vibrator
	MC208	20-25W	6/12	single-unit	vibrator
	MC213	10W	12	single-unit	vibrator
	MC213-LP	LPI(1W)	12	single-unit	vibrator
	MC214	50W	12	single-unit	dynamotor & vibrator
	MC215	35W	12	single-unit	dynamotor & vibrator
	MC216	25-30W	12	single-unit	dynamotor & vibrator
450-470 MHz	MC306	20W	6	single-unit	vibrator
	MC316	20W	12	single-unit	vibrator

BASE STATIONS

Freq.	Number	RF Power Out.		type of control	type of mounting
152-174 MHz	SC221	50W	-	local	floor
	SC222	50W	-	remote	floor
	SC223	250W	-	local	floor (double enclosure)
	SC224	250W	-	remote	floor (double enclosure)
	SC225	----	-	local	desk (auxiliary receiver)
	SC226	50W	-	remote	pole
	SC227	50W	-	local	desk
	SC228	50W	-	remote	desk
	SC229	50W	-	local	pole
	SC230	50W	-	repeater	pole
	SC231	10W	-	local	ac mobile
	SC231-LP	LPI(3W)	-	local	ac mobile
	SC233	250W	-	local	floor
	SC234	250W	-	remote	floor
	SC241	25W	-	local	ac mobile
405-425 MHz or 450-470 MHz	SC321	20 or 30W	-	local	floor
	SC322	20 or 30W	-	remote	floor
	SC325	--------	-	local	desk (auxiliary receiver)
	SC326	20 or 30W	-	remote	pole
	SC327	20 or 30W	-	local	desk
	SC328	20 or 30W	-	remote	desk
	SC330	20 or 30W	-	repeater	pole
	SC337	20 or 30W	-	local	Desk-Mate
	SC338	20 or 30W	-	remote	Desk-Mate

to be generally available. The 6939, for example, a fairly common miniature tube for some RCA and Comco units, can't be found in the usual "used" or "reconditioned" tube markets—and the replacement cost is a fat $16. The cost of only a few at that price might pay for a complete and fully tubed rig of another make.

Timing Devices for Remote Control Applications

The common denominator of all control systems is the timer. And a multiplicity of timed functions is what gives the control scheme the character of a brain. From initiation of the first control pulse, which starts a planned sequence of events, the timers take over, commanding all the "whens" designed into the system.

Though timers fall into but three basic categories, they cover a virtually unlimited range of configurations, including capacitive, semiconductor, motor-cam, thermal, pneumatic, and clock. Each type has peculiar advantages for specific applications, and not all are ideally suited for use in repeater control applications.

The thermal timer, for example, is small and inexpensive, and it often is proportioned to plug into a standard miniature tube socket. But its reset time is dependent on its cooldown ability. Once triggered, the thermal timer cannot reliably operate at the same period until its thermostatic element reduces to a temperature comparable to its original state—which may be many minutes. In applications where the timed sequence must be accurate, and where there is a chance the timed cycle must be repeated at short intervals, the thermal timer is a complete bust.

The purpose of this article is to outline the basic timer functions and point out a few of the principal operating characteristics of the most-used types for repeater control. Using this data, designers of control schemes can select the most advantageous timer for a particular application consistent with funds and skills available and with degree of accuracy required.

TIMER FUNCTIONS

The three timer types are the delayed dropout, the time delay, and the interval timer. Figure 1 shows the three functions graphically, in terms of voltages being controlled by the timers.

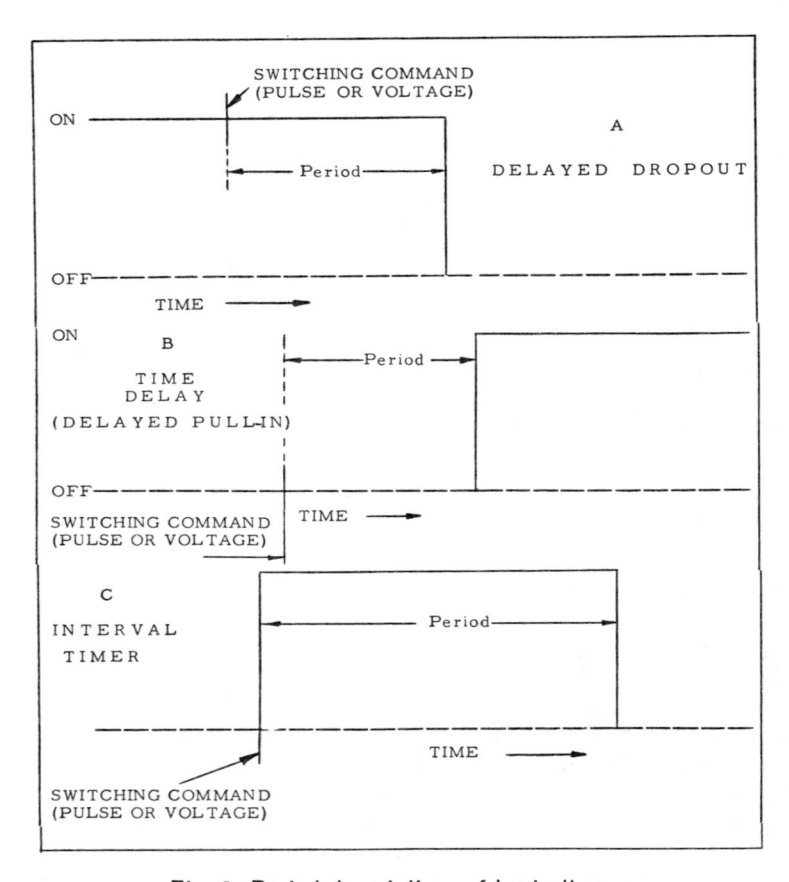

Fig. 1. Period descriptions of basic timers.

The overall character of any timed sequence can be shaped according to individual requirements with the aid of conventional switching functions (solid-state devices or relays).

Some timing devices can be difficult to classify, because their functions vary according to the control voltage application. The Agastat* line of pneumatic delayed-dropout timers is a case in point. This device turns on as if it were a conventional relay, and remains in this state until coil voltage is removed, at which time its period starts. At the end of its period, the contacts release and the timer is "off." Use of a pulse, rather than a continuous voltage, however, gives the

---

*Registered Trademark, Elastic Stopnut Company, Elizabeth, New Jersey.

Agastat delayed dropout the character of an interval timer, in that a specific and precise "on" period results from each pulse application.

The most widely used delayed dropout timer is the simple capacitive device, consisting of no more than a relay whose coil is shunted with an appropriately selected electrolytic capacitor. This type of dropout cannot be triggered with a short pulse because the electrolytic must be fully charged before the timed function can occur. And adequate charging, in some cases, can take several minutes. The tendency of an electrolytic to resist voltage changes is a characteristic which can be used to turn the delayed dropout into a conventional time delay as well. The requirement here is to reduce the coil voltage with a resistor to just that amount required to energize the relay—no more. The capacitor, then, when placed across the coil, opposes the voltage and keeps it below the critical triggering point until it is effectively charged. (See Fig. 2.) The limitations of the capacitor-relay timer are more numerous than serious, but often these limitations are of such a nature as to prevent the use of this type of timer in an otherwise relatively uncomplicated switching scheme. The period, for instance, may vary considerably, dependent on such local conditions as ambient temperature and voltage stability. Timers of this class should only be used where energization time is not critical and where the timed period can vary a great deal without deleteriously affecting system operation.

TIME DELAY (DELAYED PULL-IN)
Relay coil at A sees gradually increasing voltage which eventually reaches "critical" (switching) level. Resistor reduces voltage to just that amount required to energize relay.

DELAYED DROPOUT
Relay coil is supplied with sufficient voltage to provide fast pull-in of contacts. After capacitor reaches full state of charge it will hold coil energized for a time after coil voltage is removed.

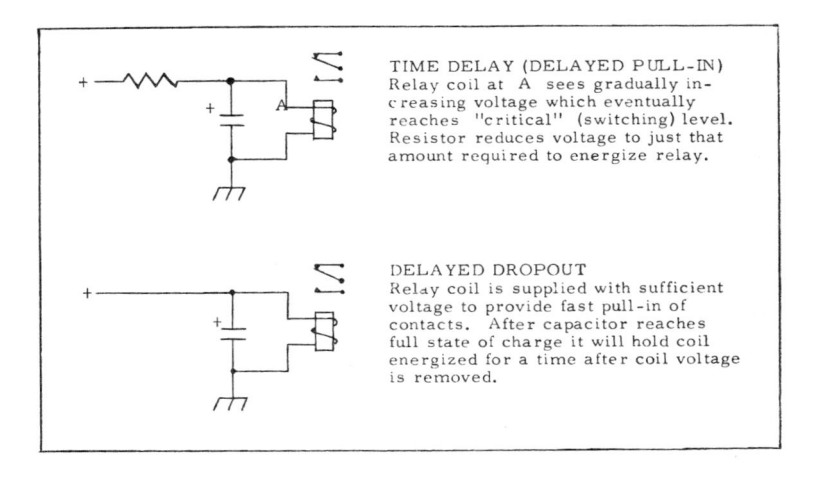

Fig. 2. Use of relays and capacitors to control uncritical timed switching sequences.

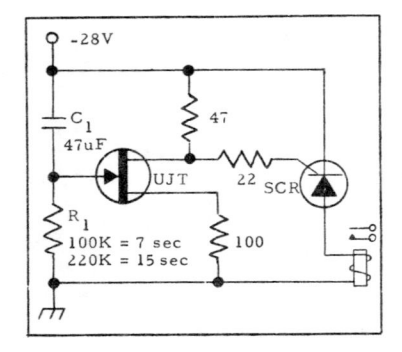

Fig. 3. Solid-state interval timer (pulse or voltage keyed).

Solid-state timers—particularly those employing SCR's—provide an ideal means of timed sequencing in situations calling for reliability, positive switching, accuracy of period, and ease of triggering. The only significant disadvantage with solid-state timing devices is their inherent complexity. Often a control system will require as many as five discrete timed functions. With a solid-state scheme, a separate and complete electronic circuit is required for each timer.

Figure 3 shows a proved circuit for a solid-state interval timer capable of being actuated by either a pulse or a continuous voltage. The timed period, varied by selection of the potentiometer (R1), may be anywhere from 0 seconds (zero resistance) to 15 seconds (approximately 220K). The period can be upped to several minutes by using a 100K resistor for R1 and increasing the value of C1.

Although the advantages of this solid-state interval timer are obvious in terms of dependability, period repeatability, and general imperviousness to environmental conditions, one can readily see the drawbacks where a large number may be required in a given control application.

The Agastat line of time delay relays is particularly suitable for repeater applications because of a favorable compromise in accuracy, economy, reliability, and convenience. The Agastat is available as a delayed dropout or delayed pull-in; it is small, compact, and completely self-contained; and it is as easy to use as a conventional relay. In most cases, the Agastat is provided with an integral vernier adjustment for varying the time (often from 75 percent of its preset period to several hundred percent).

The Agastat is a pneumatic timer, whose period is determined by air pressure against an integral diaphragm. In the case of the delayed pull-in, pressure begins to build up with application of coil voltage. (The pressure point of switching is determined by diaphragm position which is adjusted by the vernier.) When the pull-in takes place, the timer will remain energized in exactly the same capacity as a relay until coil voltage is removed, at which time it will immediately drop out and will be ready for the next timed sequence.

Factory fresh Agastats are out of the price range of most amateurs, unfortunately. But an abundance of them on surplus shelves brings the acquisition cost down to a level comparable to that of quality thermal timers—about $5 each.

It is easy for a system designer to end up with an excess of components in his control system. Careful study of timers and their characteristics will help to minimize this eventuality. My own telephone control circuit was an example of overcomplication with timers. Figure 4a illustrates the approach presented in the original article for timed automatic telephone shutdown. Inversion of the design logic, however, simplifies the circuit by eliminating a relay (Fig. 4b).

Both sketches in Fig. 4 show the accomplishment of a single function. The extra relay in 4a is the result of constricted thinking. When a carrier appears, the COR pulls in to provide a signal at K1 to prevent the timer's operation. In 4b, the COR provides a signal when it is **not energized**, which can feed the timer directly. In both cases, the timer pulls in to turn off the control voltage when no carrier occupies the input frequency for a specific period.

My mistake in the original design was in thinking "I don't want X function to happen while the carrier is present." This line of logic led me to insert a control element (K1) to deenergize X function in this condition. A more positive approach would have been "I do want X function to happen when no carrier is present." This statement immediately suggests direct energization of X function from the no-signal state.

If your own control system is still in its design stage, look again at your timed sequences. Proper selection of timing devices can save you a lot of unnecessary components. And bear in mind that complete control systems can be constructed with timers as the sole function selection mechanism.

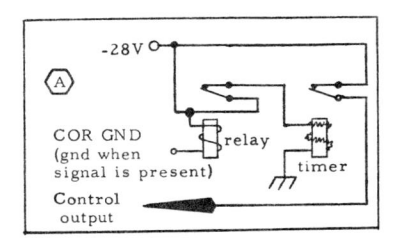

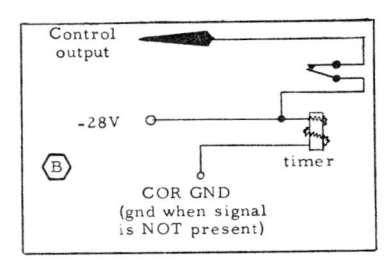

Fig. 4. Logic inversion to simplify control scheme.

The Two-Way Repeater

by Gordon Pugh W2GHR

Have you ever thought that a "local" repeater—serving a limited area for that everyday communication—should also be capable of "extended" coverage? Sounds easy—and in some cases, it is. A different input channel or perhaps tone coding could just key up the PA.

Here is an approach to the local-and-extended coverage operation when the repeater is at the edge of the "local" service area using directional antennas to limit access and coverage to a specific area.

The repeater site in question is on the edge of a large metropolitan area where there is considerable activity on the input frequency. The circuit to be described keeps an ear on the overall input channel activity but allows weak stations in the "local" area to capture the station in the presence of strong signals from outside. The first requirement is similar coverage of the local area on both the local and omni antennas. The repeater must be capable of operating on either antenna without receiver or transmitter adjustment or interference. The only requirement is a quick-release COR (by adjustment, in most cases) and a good quality antenna relay (it will see many operations under no-load conditions).

The circuit will accept tone-coded (tone burst) signals on the omni antenna and all signals on the local antenna. This allows "extended" operation from either the local area or from within the extended area upon request (tone burst) by the user. At the end of each continuous transmission the station returns to the local-open / extended-tone coded mode and will examine each incoming signal received.

Circuit Description

A single receiver is used with a carrier-operated relay (COR) and audio tone decoder. Relay closures from the COR and tone receiver are connected to the selection circuit. A high-quality coaxial relay is connected to the diplexer "antenna" terminal, the normally closed path connects to the omni antenna and the normally open terminal connects to the directional antenna.

Figure 1

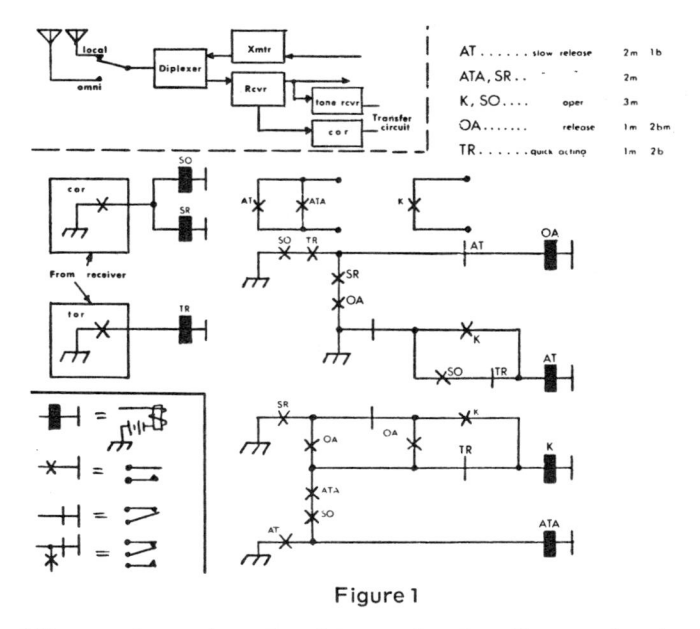

When an incoming signal is received on the omni antenna, contact COR closes a path to the SO and SR relays. The SR operates quickly. The SO relay operates after 150 ms. If a tone is present at the tone receiver, contact TOR operates the TR relay. A path is completed through the TR contacts and the SO contacts operating the OA relay. The OA relay locks on itself through the normally closed AT contacts, the SR and the OA contacts. Closure of the OA relay opens the path to the AT relay and closes a path through the OA and SR contacts in the K relay circuit. The operated TR relay prevents operation of the K relay. When the tone ends the TR releases, completing the circuit and operating the K relay which locks out the TR contacts, preventing false release of the K relay by subsequent tone signals. The K relay operates the repeater input keying circuits to activate the repeater. When the COR releases, the SO releases followed by the SR, the K and the OA relays.

When an incoming signal without the tone is received on the omni antenna only the COR operates, closing a path to the SO and SR relays. Operation of the SO relay closes a path through the normally closed contacts of the OA and TR relays and the SO contacts operating the AT relay. Operation of the AT relay opens the path to the OA relay, closes the path to the ATA relay, and operates the antenna relay, switching the receiver to the directional antenna. Operation of the ATA

227

relay follows the AT completing the operation path to the K relay through the normally closed TR contacts, the ATA, AT, and SO contacts. Relay K is slow to operate and pulls in only after time has elapsed to permit the COR to drop out, releasing the SO relay and breaking the K relay operation circuit. If the COR drops out due to lack of signal at the directional antenna, the AT relay will release after 250 ms, breaking the ATA relay operating circuit. During this interval the antenna relay remains operated holding the receiver on the directional antenna. With the release of the ATA relay 250 ms after the release of the AT, the antenna relay releases, returning the receiver to omnidirectional monitoring.

If a signal is again received, the sequence is repeated.

If the COR remains operated after transfer to the directional antenna, the circuit is completed to the K relay which energizes, locking the AT relay until release of the K relay and transferring control of the K relay to the SR relay. Momentary release of the COR relay is not transferred to the K relay because of the SR relay action.

The Second Input Channel

by Gordon Pugh K2GHR

In the northeast, as in some other areas of the United States, repeaters are faced with a serious problem—too much coverage. New repeater installations will, in many cases, have coverage that overlaps the existing areas of other repeaters. Unfortunately, it is seldom possible to tailor the service area to avoid this overlap.

Most repeaters in the northeast, with the exception of the Washington, D.C. to New York City area, operate by repeating 146.34 to 146.94 MHz. As long as the coverage areas do not overlap—or do so in lightly used places—few problems will be encountered. However, if two or more repeaters using the same input channel serve an area of considerable activity, chaos may result. Unusual band openings have produced this condition occasionally in recent years.

Selecting the desired repeater becomes a problem. There are several avenues to reduce unintentional operation of more than one repeater:

1. Use tone coded selection of each co-input-channel repeater.
2. Use directional antennas to reduce the signal radiated toward the unwanted repeater.
3. Use an alternate input channel at each repeater.

Many areas have run out of frequencies that could be used as second channels. It would be worthwhile for all repeater users / operators in a region to coordinate frequencies in order to obtain the best use of both the 60 kHz and "split" channels.

The second channel is equally effective in areas where the repeater output is on 146.76 MHz or where the output channels alternate between 146.76 and 146.94 MHz. While the second channel will not eliminate overlapping transmitter coverage of adjacent repeaters, it can substantially reduce interference due to keying of unwanted repeaters.

One might well ask the question, "Must I carry a whole trunkload of crystals to use these repeaters?" The "second channel" concept does indeed tax the capacity of most mobiles. And this is an unfortunate condition we simply have to live with. Base station operation, where the second channel does the most good, is more easily equipped for multiple frequencies. But the mobile situation isn't entirely hopeless: Maybe some of the eleven-channel mobile sets will become available in the near future. Or better yet, maybe an inexpensive, simple, and totally compatible synthesizer will come along. The outlook in this regard seems the most promising.

It must also be compatible with the rest of the station operation. If a duplexer is used at the repeater, the maximum separation between the two input channels is about 60 kHz with the duplexer tuned to the midpoint. The normal insertion loss of 1.6 dB for the Sinclair F150-4E duplexer increases to about 2 dB when used this way. Repeaters using remote receiver sites or other isolating methods are not as severely limited but should keep within the limits of any antenna coupler in the receiver circuits. Mt. Mansfield (W1KOO) and Mt. Beacon (W2CVT) are not setting up second input channels on 146.37 MHz. The land use application for Mt. Snow includes a second input on 146.31 MHz. A second input will be available to Concord, N.H. from Ascutney, Vermont, on 146.46, tone-coded. And the Lenox Mountain repeater group has selected 146.22 MHz as a second input.

The second input has one other advantage. When two or more repeaters can be interconnected on frequencies above 220 MHz, second-channel receivers can be set up at adjacent repeater sites without interfering with the local operation at either repeater. For example, a remote receiver on 146.28 MHz can be installed at Mt. Mansfield and connected to Killington on 220 MHz, and a remote 136.37 MHz receiver can be installed at Killington and connected to Mt. Mansfield on 420 MHz. In this case, the remote receiver at each site would be operated from the common antenna and duplexer without interference to the existing station. Since the remote second-channel receivers will both be 60 kHz off the center frequency of the duplexer at these sites, some additional loss will be introduced. This loss is desirable when the overall site is considered.

Tone coding will work as a means to select one of several repeaters, though it will not solve the main channel congestion problem. It appears that tone coding will eventually be necessary on a common input channel if activity continues to increase.

Directional antennas and limited power would aid overall operation as has been demonstrated in northern Vermont. The directional pattern reduces congestion on the common channel. This type of installation is desirable at a fixed station but is quite impractical in a mobile.

The alternate-channel approach has more to offer. If regional coordination of second or alternate channel inputs is established, each repeater would have an input channel that would not carry traffic for another repeater within the reception limits of the station. The second input would reduce congestion on the main channel, provide access to only one repeater on an "uncluttered" frequency (that could remain open or uncoded), and also permit remote receiver installations at adjacent repeaters without mutual interference. By using the second channel for base station input, a mobile station could break in more easily when operating in the repeater "fringe area" if both inputs were usable simultaneously. Duplex operation would also be possible.

Repeater activity has expanded north along the Hudson and Champlain valleys during the past ten years. Stations were installed in New York at Mt. Beacon, Troy, Schenectady, and in Vermont at Mt. Mansfield. All the repeaters transmit on 146.94 MHz and receive on 146.34 MHz (except Schenectady, which receives on 146.46).

Two of the repeaters, Schenectady and Mt. Mansfield, are horizontally polarized. The Schenectady and Troy repeaters share much of the same area but are cross polarized and are on different input channels. The only other overlap has been between Mt. Beacon and Troy in a lightly used area of the Hudson Valley. There were many areas of no coverage between the repeaters, including central-southern Vermont and the western portion of Massachusetts and Connecticut.

A repeater was set up at Killington, the highest mountain in central Vermont, to fill the gap between Troy, Mt. Mansfield, and a repeater in Concord, New Hampshire, also operating on 146.34 to 146.94 MHz. In order to cover this area, it also overlaps the other coverage areas. Mobile stations using these repeaters occasionally pick up Killington. Many base stations operating on 146.34 MHz key Killington in addition to the intended repeater. Recently, W1KOO converted to vertical polarization and many of the stations using W1KOO also changed to vertical. Following the change, about 90 percent of the traffic through W1KOO was also transmitted by W1ABI at Killington.

This overlap had been predicted and the repeater at Killington was designed to repeat either 146.34 or 146.28 to 146.94 or 146.88 independently. In addition, the 146.34 receiver can be switched to either "open" or tone-coded operation. If it is programed to transmit on both outputs, selected by the input channel, use may be made of the alternate output frequency on a time-shared basis with normal operation. Break-in is still possible.

Selection of a second input channel should be carefully coordinated. The frequency should not be reused within twice the maximum normal base station working range into the repeater.

Touchtone Control
by Gene Mitchell K3DSM

Touchtone is becoming ever more popular in amateur radio remote signaling applications. Touchtone dials manufactured by Western Electric and Stromberg Carlson have already been seen on the surplus market. These range from the 10- to 12- and even 16-button touchtone dials used in military and computer application.

Two tones are generated when a digit button is depressed. The output tones pass over the same leads as the supply

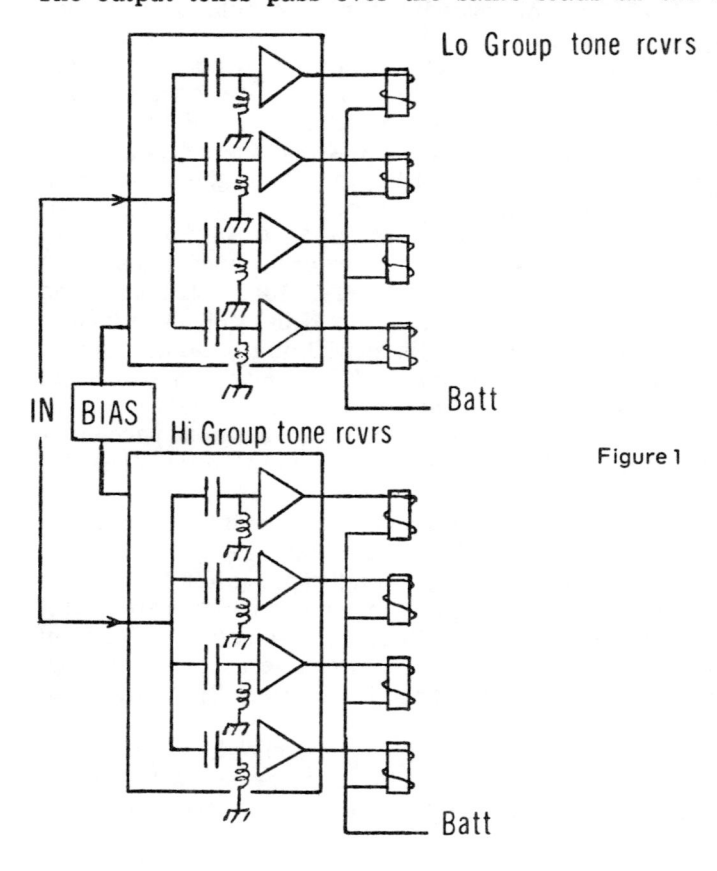

Figure 1

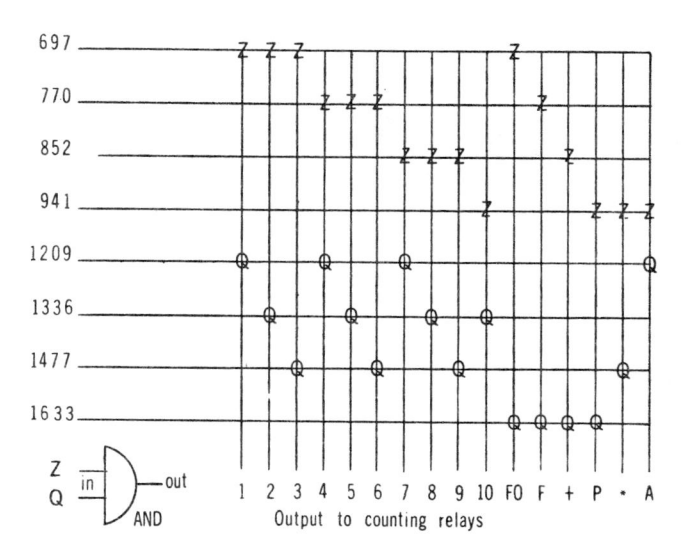

Fig. 2. Solid-state logic circuit.

voltage, similar to the way transistor microphones work. The dial can be connected to the input of many transmitters using carbon mike input. Sometimes the supply voltage drops under load, but minor changes will permit their use directly into the transmitter. Supply voltage can be around 6-12 volts across the dial.

The block diagram of the tone receiver shown in Fig. 1 is simplified from those used by military, computer, and telephone applications. Amateur radio application is not as critical. Much of the limiting takes place in the FM transmitter and receiver. Codes can be arranged so that not all tones will be needed. In this case, the tone receiver can be simplified. Part values such as relay and transistor types are unavailable, but this is no big problem since scrounging is a normal process and types may not be critical. UTC type VIC inductors are ideal since they are tunable; 88 MHz toroids may also be used by selecting proper values for capacitors.

Instead of the output relays for the tone receiver, solid state AND gates may be used. The logic is shown in Fig. 2.

Hybrid Coupling in Remote Telephones

Connecting a phone patch into a remote system so that the telephone can be answered or used for initiating dialed calls can present some interesting problems because of the circuit parameters of the phone patch itself. But before the effects of these characteristics can be understood, it is necessary to have a basic understanding of the workings of a conventional telephone.

A telephone line typically consists of three wires. One of these is a floating ground, and need not be considered with respect to the actual circuit. The remaining two wires (L1 and L2) carry all signals for both ringing and communication.

When the telephone rests on its cradle, the line (L1 and L2) is essentially an open circuit; at least its impedance is high enough to appear open. When the phone rings, a fairly high ac voltage is developed across the open lines. The high voltage ac is coupled to the bell inside the telephone instrument as long as the handset is in place. When the handset is lifted, however, an electrical short is placed across the line (in the form of an inductor). As long as a dc circuit exists between L1 and L2, the telephone will be "off the hook," but if the circuit is broken for any length of time, a "hang-up" will occur. If it is broken at rapid intervals, the act of "dialing" takes place.

Thus, for ringing, L1 and L2 must be isolated; and for communicating, they must be shorted (with a 600-ohm inductance).

As it happens, typical phone patches offer an isolated L1-L2 pair internally. The manufacturers of these patches have designed the units for use by individuals who will physically remove the handset from its cradle. If the L1-L2 pair were not internally isolated, amateurs would be unable to leave the patch in the phone circuit.

When a phone patch is used in remote radio applications, it **cannot** have an internal isolated pair. If it did, the telephone would go right on ringing after the answering circuit has been energized. A remote patch **can**, however, have a dc internal L1-L2 pair. The reason for this is that the phone patch is not a part of the telephone circuit **until** the remote operator sends the "phone on" command. Therefore, the isolated L1-L2 pair of the phone line is used for ringing when the "phone on" circuit is not energized. At energization, the phone patch can be used to "short" the line.

Here is where the selection of the right patch is important. Many of the less expensive units contain a pair of series transformers separated by a capacitor. Shorting of the capacitor may provide the dc path from L1 to L2, but if the transformer windings present a mismatch, serious gain loss can result.

Three commercial patching units were tested for optimum performance in the author's remote telephone system: the Monarch, the Johnson, and the Waters Universal Hybrid Coupler. The Monarch exhibited a gain-loss characteristic to such an extent that no amount of control manipulation could remedy it. The Johnson patch exhibited a moderate gain loss, but not so bad it couldn't be tolerated. The audio null in the Johnson phone patch was also not quite what one could consider optimum, either, although it was by no means unacceptable. Good transformers in a good phone patch design should give sufficient null so that receive audio does not get fed back into the transmitter through the phone patch coupling circuitry. This is doubly important where the repeater already couples receive audio to the transmitter. Interconnection of the hybrid patch should have minimal effect on overall system audio. Insufficient null greatly increases normal repeater audio output without commensurately increasing the incoming audio from the telephone to the transmitter.

The Waters Universal Hybrid Coupler performed the best (which was to be expected, since it is the most expensive of the three). The Waters unit provided several bonus features, too, that proved very beneficial for autopatch operation.

First, getting a proper dc continuity on the Waters unit was simplified because the coupler contains an internal 600-ohm transformer across the line (for making tape recordings). Figure 1 illustrates this portion of the circuit.

As can be seen from the schematic, a dc circuit in the Waters coupler can be obtained by merely lifting the top leg of the transformer primary winding (yellow lead) and moving it to the other side of the capacitor.

Another bonus feature of the Waters coupler is its built-in compressor-amplifier, which tends to normalize all telephone signals to a constant value. The repeater transmitter is thus supplied with a uniform level of audio from the coupler,

orig circuit

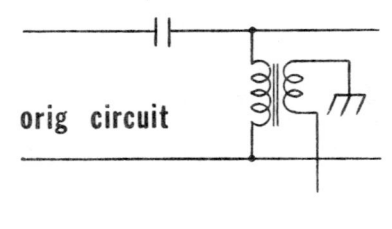

modified circuit

Fig. 1. Waters coupler and Autopatch mod.

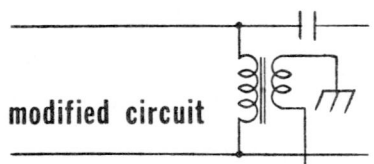

regardless of the audio characteristics of a particular telephone circuit.

But the most noteworthy aspect of the Waters coupler is its null capability. In the autopatch installation in which the coupler was used, sufficient null was achieved so that repeater audio remained the same whether the phone patch was in the circuit or not, and the audio from the phone line could be set to the same level as the incoming receiver audio. Reports from parties on the landline end of the phone patch proved that no significant differences existed between the automatic patch and a conventional telephone connection.

A simplified version of the circuit in which the phone patches were tested is shown in Fig. 2. That portion of the

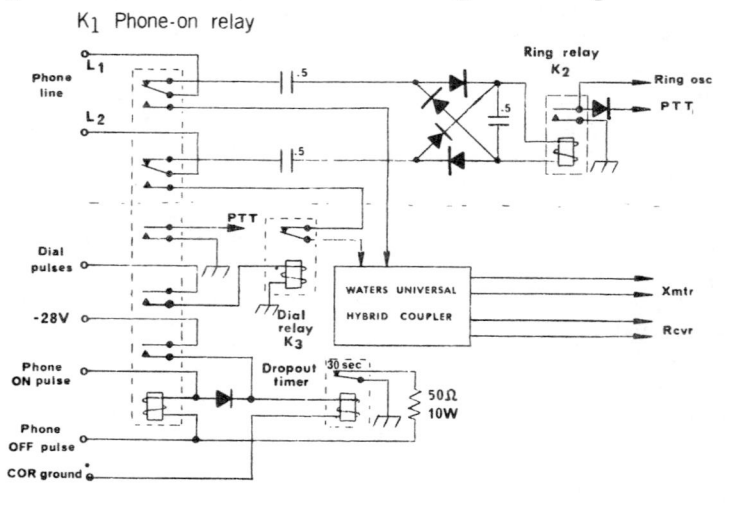

•Ground when signal is NOT present
All pulses -28V

Fig. 2. Simple Autopatch circuit.

Photo shows waters coupler (left), timers, and control relays. Pictured autopatch contains additional access circuitry not shown on schematic (for such "extras" as times on, timed off, cor-energization through anding, and automatic shutdown for excessively long transmissions).

circuit above the horizontal broken line is the "ring" circuitry. When the ac "ring" voltage appears on the phone line, it is rectified and used to key a sensitive plate relay (5-8K ohms). When the ring relay pulls in it keys the transmitter push-to-talk and triggers a simple audio oscillator connected to the transmitter mike line. The diode is used to keep the push-to-talk circuit from energizing the oscillator each time the carrier-operated relay is keyed.

When the phone rings, or when the remote operator wishes to place a call, he transmits a signal that will give him a low-voltage negative pulse from his own control circuit, which pulls in and latches the "phone-on" relay. (The 50V diode keeps the "on" pulse from being overworked.)

The phone-on relay accomplishes the task of lifting the receiver from the hook, and couples the L1-L2 line into the dc circuit of the patch. Dialing, then, is simply a matter of rapidly opening and closing the dc circuit, which is done by pulsing the dialer relay with low voltage signals from the repeater's control system tone decoder.

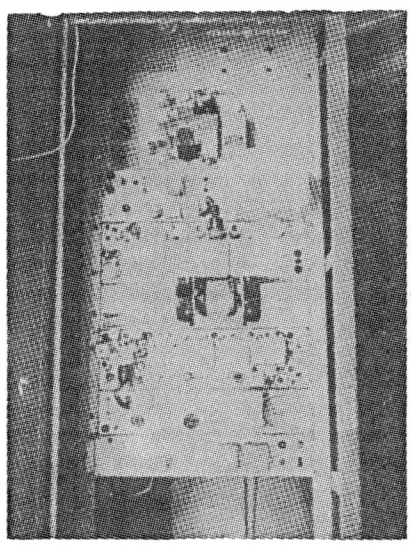

Autopatch (top panel) in combination with conventional 450 MHz repeater and a limited control system provides users with Handie-Talkie and mobile access to telephone line.

In the circuit pictured, hang-up is achieved by merely releasing the phone-on relay. This can be accomplished both actively and passively. For active shutdown, a low-voltage negative pulse from the control system will place a voltage on the ground side of the phone-on relay which is of the same polarity as the opposite coil terminal. The 25-ohm resistor accepts the load during the pulse and the relay is released. Passive shutdown is done with a 30-second timer whose period begins the instant an input carrier disappears. When a signal comes on the repeater input frequency, the timer is defeated, and the period will begin anew when the carrier-operated relay is released.

An automatic phone patch can be as simple or as complex as the designer wants it to be—and the only difference is the ease of access. The performance of the system, however, is dependent almost entirely on the selection of a suitable audio coupler. A good generalization, then, would be to advocate skimping in everything but the audio coupling unit—but here, if you're going to buy, by all means, buy the best!

Practical Circuit Applications Using that Strange Diode: the Varactor

by Bill Mengel

The varactor is a simple two-terminal device extending dependable operation in the VHF, UHF, as well as microwave frequencies by utilizing the voltage-variable capacitance of a pn junction. The varactor provides a way of tuning circuits, multiplying and dividing frequencies, controlling frequencies, and performing other functions. A varactor, which is a special-purpose junction diode, has been designed to make its junction capacitance useful; it is because of this property of a varactor that capacitance, which is an unavoidable nuisance in conventional diodes, is purposely cultivated into the varactor. The basic configuration for the varactor is shown in the illustration below.

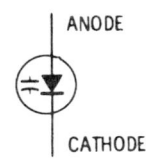

The operating portion of a varactor is in the region where a conventional diode would be considered to be cut off—principally in the region between forward conduction and reverse breakdown. In most cases, the varactor is reverse-biased since in this state it draws a minimum of current, making it essentially voltage-operated. The behavior of the pn junction of the varactor at different applied bias potentials is as follows:

ZERO BIAS— At zero bias, the contact potential is determined by the semiconductor. There is no change in capacitance and no current flowing at this time.

FORWARD BIAS—When forward-biased, high forward current flows as the external voltage applied is in series with the contact potential. The contact potential decreases thus increasing the capacitance.

REVERSE BIAS—When reverse-biased, the external voltage applied is in parallel with the contact potential. The contact potential increases, extremely low reverse current flows, and the capacity decreases.

The property of being able to vary the capacitance by changing the applied voltage enables the varactor to do the work of a conventional variable capacitor many times its size. The capacitance of a varactor varies inversely as the reverse voltage, and directly as the forward voltage. It may also be noted that the capacitance of a varactor also varies nonlinearly. Varactors also have a Q approaching that of air trimmer capacitors, so they could be used in such locations as rf front ends and high-efficiency multipliers as well as other normally sensitive circuits.

The varactor diode by itself is unique in frequency multiplying and dividing. First, the rf signal itself is the only power required to operate the varactor. Secondly, the varactor, by distorting the input signal develops an output rich in harmonics. Thirdly, a varactor can provide a means of high power output at frequencies normally beyond the limits of present power transistors. In frequency multiplication, it is only a matter of placing a tuned circuit (tuned to the input frequency) on one side of the varactor and placing another tuned circuit on the other side tuned at the desired harmonic. As shown in Fig. 1, the input circuit is tuned to frequency f. The output of this circuit is then fed to the varactor where it is distorted. This distorted output is then fed into an output circuit tuned to frequency f(n) out.

In typical frequency doublers, efficiencies as high as 90 percent—as compared to the 50 percent efficiency of conventional tubes and transistors—can be realized. This can

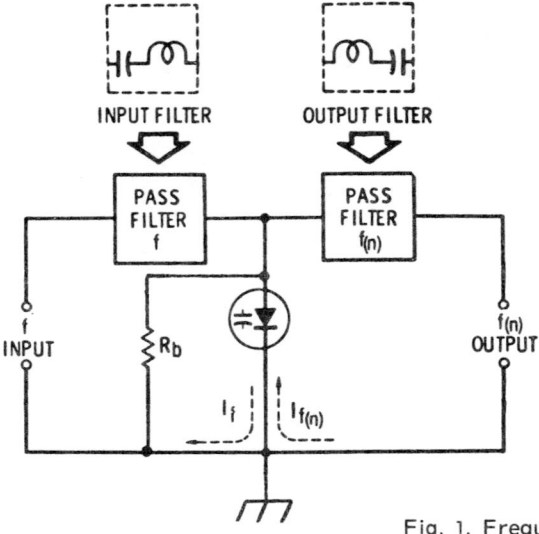

Fig. 1. Frequency multiplication.

Fig. 2. Postdoubler multiplication.

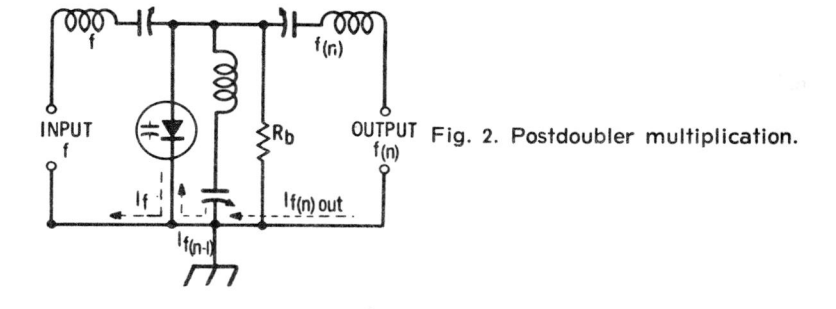

be attributed to the fact that a varactor dissipates very little power and has low loss. A properly designed varactor multiplier does not generate noise. However, parametric oscillations can occur from highly overdriven varactors or from unwanted idler resonances. A bias resistor R_b (shown in Fig. 2) will usually have a value of from 68K to 270K. The higher values of resistance make the circuit more efficient while the lower values of resistance make the circuit operate more linearly.

Since average capacitance varies with input power applied, some detuning will occur if the input power to a multiplier using a varactor is changed appreciably. All frequency multipliers beyond a doubler require an idler circuit for maximum efficiency. An idler circuit is used to reinforce the output frequency of a multiplier. This is done in the following manner. The current developed by the idler circuit is added to the fundamental current to form the harmonic current. The tuned frequency of an idler is generally set to one harmonic below the output frequency, as illustrated in Fig. 2.

A basic example will now illustrate the principles of operation of a typical varactor circuit. Our problem is that we want to take a present signal of, say, 150 MHz and develop an output of 450 MHz.

Referencing Fig. 3, capacitors C_1 and C_2 are used to match the input and output of the tripler to the input and output impedances. With an input frequency of 150 MHz, the input filter is tuned to a frequency of 150 MHz. A type 1N4387

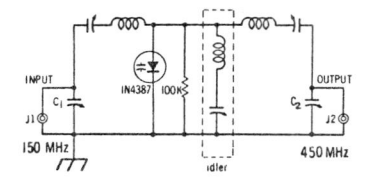

Fig. 3. Varactor tripler circuit.

varactor is chosen. This varactor is capable of 60 percent efficiency at 450 MHz (offering a power output of 18 watts with an input of 30 watts). The idler circuit is tuned to one harmonic below the output frequency. In this case the idler should be tuned to resonate at 300 MHz. The bias resistor is chosen as 100K so the circuit will operate linearly. The output circuit is then tuned to resonate at the desired output frequency (450 MHz). After alignment, it is a good idea to repeat the tuning procedure because there is almost always some interaction between stages.

Another use for the varactor is in the development of an FM signal. By rectifying a modulated signal and applying that fluctuating voltage to the terminals of a varactor, we could use the changing capacitance of the varactor to cause frequency deviation of an oscillator. Hence, the development of frequency modulation via the varactor. Also, by properly proportioning the fluctuating audio voltage going into the varactor with respect to the oscillator, either narrowband or wideband FM may be obtained, as shown in Fig. 4.

In the circuit of Fig. 4, a rectified audio voltage is introduced at the potentiometer which can be adjusted to allow the required frequency deviation whether it be wideband or narrowband. The charging and discharging of capacitor C1 through resistor R1 applies a fluctuating voltage on the anode of the varactor. This fluctuating voltage will cause the capacitance capabilities of the varactor to vary, thereby pulling the oscillator off its center frequency.

As mentioned earlier, the property of being able to vary the capacitance of a varactor by varying the input voltage enables it to do the work of a conventional variable capacitor. One great advantage as opposed to conventional tuning is miniaturization. A typical varactor for this type of service in

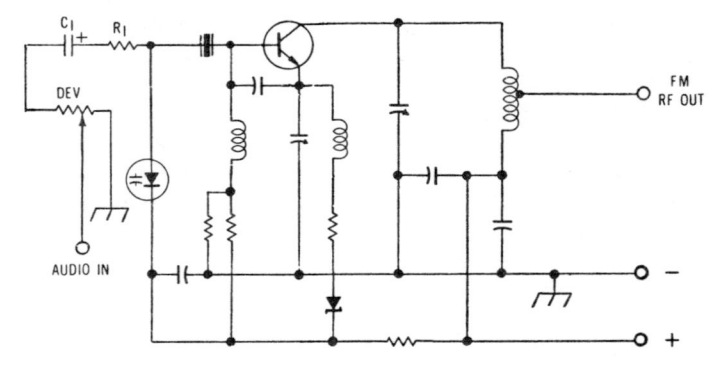

Fig. 4. Frequency modulation using the varactor.

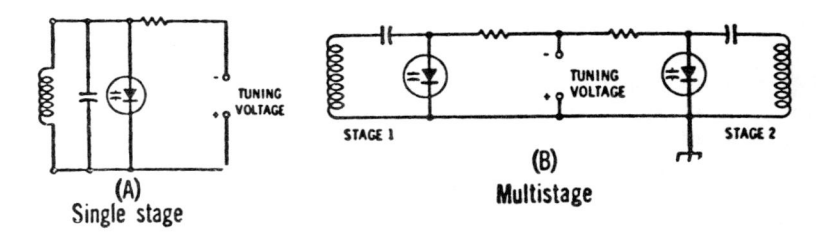

Fig. 5. Varactor tuning.

most cases is about the size of a small signal diode (1N34, for example) and this is many times smaller than even the smallest variable tuning capacitor. In cases where larger values of varactors are needed than is available, parallel operation is feasible. However, it must be kept in mind that both the minimum and the maximum capacitance capabilities are increased with parallel operation. Multistage tuning that at one time required a large ganged variable capacitor can now be controlled by a single small variable potentiometer by varying the dc control voltage to the varactor. The illustrations of Fig. 5 show a typical circuit using a varactor for tuning along with a circuit utilizing varactors for multistage tuning.

In the case of an FM receiver, a varactor can be utilized to regulate the amount of drift of the local oscillator by com-

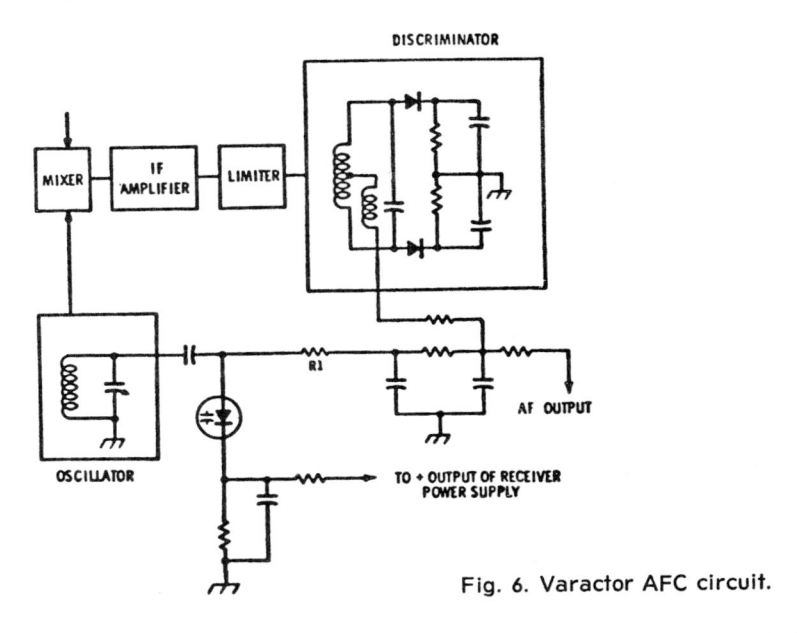

Fig. 6. Varactor AFC circuit.

243

pensating for that drift and, in a sense, locking it on frequency. This type of circuit is commonly known as automatic frequency control or simply AFC.

What occurs in a typical AFC circuit (Fig. 6) is this: A correction voltage developed in the discriminator circuit is directed to a varactor through a filtering network. Any error in tuning will result in a voltage change at the discriminator and it is this change that is used to alter the capacitance of the varactor to compensate for that error. This changing capacitance is then used to complement the final tuning of the oscillator to lock it on frequency.

The possibilities of a varactor in communications applications are almost limitless. Scan-tuning, a technique that once required many complicated circuits, is now simplified by a varactor: With scan-tuning, band sweeping is accomplished by applying a fluctuating voltage from a sawtooth oscillator. The sweeping rate is then predetermined by the frequency of that sawtooth oscillator.

This just briefly illustrates how the varactor, a comparative newcomer to the field of semiconductors, opens the door to simplifying and improving many different types of electronic circuits.

Multiplex

by Gordon Pugh W2GHR

Most repeater systems are developed around a remotely controlled base station located some distance from the control point or points. The control link is either wireline or radio, depending upon the cost of the wire facility versus that of the radio circuit. It frequently becomes desirable to transmit several individual voice signals over the same wire or radio circuit simultaneously without conflict with each other. This is known as multiplexing.

Multiplexing can be done on radio channels and over some wire facilities, depending upon the type of leased circuit. Most wirelines used in amateur remote control are short and within the area of a single central office. If the wireline is a physical copper circuit (that is, one that consists basically of a pair of wires between the two points) without amplifiers, it is usually possible to transmit several voice signals over the same wireline. It is necessary to use different channel frequencies on the multiplex to provide two-way transmission when using a single pair of wires. It may be of interest to note that the telephone companies are now using an FM multiplex system to provide additional voice circuits in places where additional cable pairs are not available. The multiplex channel operates at about 24 kHz in one direction and 60 kHz in the other. This is coupled to an existing telephone line with special filters.

There are two types of multiplexing. Frequency division is the type used in stereophonic FM broadcasting and will be considered. The other type, time division, can be used only with signals that have been converted into pulses. Pulse modulation requires bandwidth beyond the capability of equipment now in general use and also is not authorized on the VHF and lower UHF amateur bands. One item of surplus military equipment that may be modified to operate on amateur microwave frequencies is the AN/TRC-6, using pulse-position multiplex to derive eight voice-grade channels. The bandwidth required for these eight channels is several megahertz compared with less than 35 kilohertz required for eight voice channels multiplexed using frequency division SSB channels.

FREQUENCY DIVISION MULTIPLEXING

An FM or AM DSB or SSB signal may be transmitted in a certain carrier frequency band or channel in the same way that radio signals are produced. For example, a 3 kHz voice channel can be modulated on a 60 kHz radio carrier. Another signal can be similarly modulated on a neighboring radio carrier, say at 50 kHz. If these signals are transmitted over a wire or radio channel to a receiving point, they may be separated by filters and detected individually. This is known as frequency division multiplexing.

Telephone lines are generally leased as a particular grade line. Most amateur installations use the radio tieline or even a telegraph loop. (It may be interesting to note that telegraph loops in a single central office are frequently nothing more than a pair of wires that come much cheaper than the radio tieline if the distance between terminals is over 2½ miles.) The number of channels that can be transmitted over a pair of wires depends upon the type of modulation and loss of the circuit. SSB multiplex allows the maximum number of channels in the least amount of spectrum. The channels must be transmitted so that the total energy of all channels operating simultaneously does not exceed a level established by the common carrier. Loss and noise at the receive terminal will limit reception of the higher frequencies and set up limits as to the number of channels that may be used.

The same holds true to a lesser extent for radio channels. A number of voice channels may be multiplexed on a standard wideband FM carrier. Holding the deviation constant and increasing the number of channels increases the maximum modulation frequency. The deviation ratio (the ratio of maximum frequency deviation to the highest frequency in a multifrequency signal) decreases with increase in system loading (number of channels). The deviation ratio is a measure of the capability of the system to override noise.

Two methods may be used to overcome the loading: the deviation may be increased and the receiver bandwidth increased appropriately, or the transmitter power may be increased, or both methods may be used. It should be noted that the deviation ratio need not be large (telephone microwave systems with many hundreds of channels operate at or near a deviation ratio of one) but the received signal must be capable of quieting the receiver at the highest modulating frequency in the system.

MULTIPLEX VERSUS SEVERAL TRANSMITTERS

The following example will point out the advantages of multiplexing. Assume that four voice channels are to be fed from four receivers at the repeater site back to the control point. Four transmitters could be used—with four antennas, cavities, duplexers, and so on—with four receivers at the other end of the circuit each feeding a monitor speaker. Alternately, a single transmitter with four channels of CF carrier could be used—with a single antenna, no filters or duplexers, and only one receiver. Each transmitter in the four-transmitter system requires 50 kHz minimum using existing 450 MHz equipment. Not counting the third and higher order products caused by the four transmitters all radiating simultaneously side-by-side, that's 200 kHz of spectrum used up!

Frequency division multiplex can squeeze four channels of audio into about 12 kHz using single sideband for the upper three channels and retaining the 0-3 kHz slot for an unmultiplexed voice channel. To maintain a good immunity to noise, a deviation ratio of 2½ may be used; this requires doubling the transmitter deviation (or a deviation ratio of 5 may be obtained by increasing the deviation four times). The receiver must also be modified to accept the wider deviation, and audio stages require alteration to permit modulation and demodulation amplification of the multiplex signal. The bandwidth required for the multiplex system is only 80 kHz for a 30 kHz deviation. Excluding the intermodulation products of the four-transmitter method the spectrum need is reduced 60 percent by multiplexing. Power requirements for the multiplex system are similarly reduced. The power required to operate three transmitters is eliminated and the losses in rf filtering will increase the antenna power to partly compensate for the increased deviation. A tube type requires about 50 watts—or about the same as the standby power of one low-power transmitter. One terminal is needed at both ends of the system, but they each work as sending and receiving units. The power consumed by three receivers is also eliminated.

Other modulation methods may be used to generate the multiplex signal. N carrier systems, for example, use double-sideband AM. Others use FM subcarriers. All these methods require more spectrum per voice channel, reducing the advantages described above. The single-sideband method also lends itself to easy demodulation for monitoring if necessary. An audio oscillator may be connected to the receiver audio stage to reinsert the carrier of the multiplex channel to be checked. A low-pass filter inserted at the output circuit will eliminate unwanted audio products. Adjusting the oscillator to

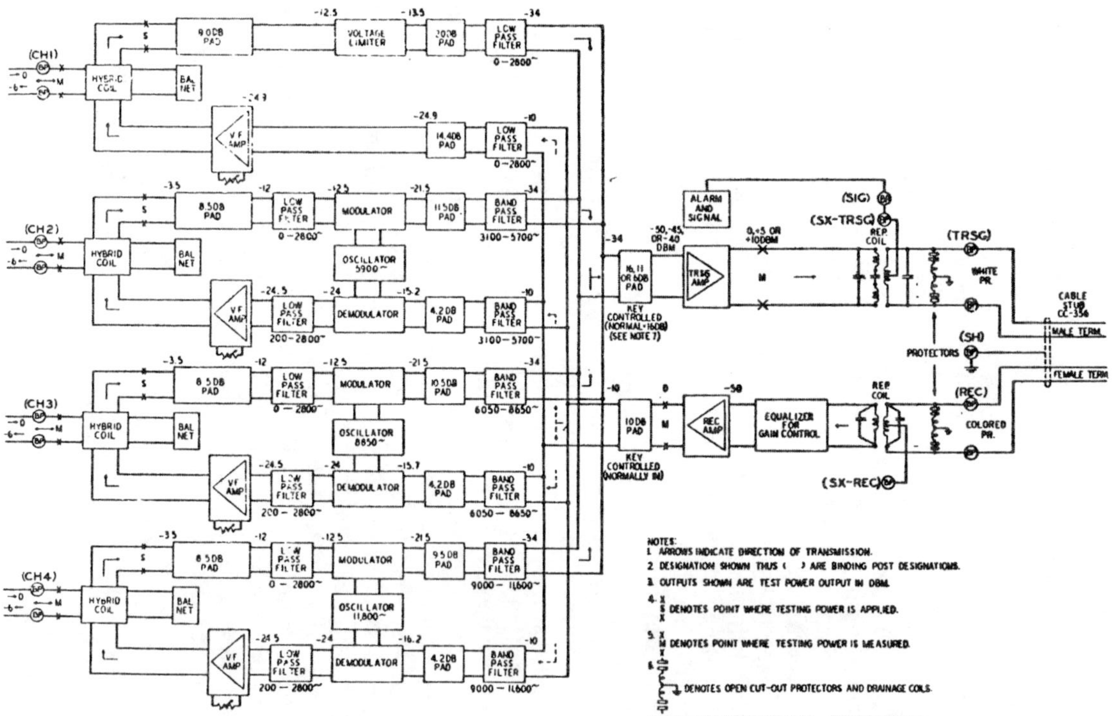

Fig. 1. Multiplex terminal unit.

the carrier frequency of the desired channel will produce the original audio information contained in the multiplex channel, provided there is some intermodulation (distortion) in the audio system being used.

The distortion may be generated in most cases by increasing the output of the audio oscillator above the linear amplification range of the audio amplifier.

Setting up a multiplex system is relatively easy if suitable equipment is available. The military carrier telephone equipment (designed for spiral-four cable) that is now considered surplus is adaptable to radio circuits. The units, called CF-1 terminals, are large and heavy but not expensive. Each terminal is installed in a six-foot rack intended for operation in the field. They operate from 115 volts ac with emergency 12V dc changeover built into the power supplies.

The CF-1 equipment passes the first channel as received except for amplification and filtering. (See Fig. 1, block diagram of terminal portion.) The other three channels are converted to A3j (lower sideband) with carrier frequencies at 5.9, 8.85, and 11.8 kHz using balanced modulators and bandpass filters for each channel. The local oscillator for each channel is used for generating and demodulating the multiplex signal. (A typical modem or channel unit is shown in Fig. 2.)

CF-1 equipment does not generate any pilots or reference signals to lock the local oscillators together or maintain constant audio level under varying conditions. Feedback control of the levels is relatively unimportant in a properly aligned FM transmission system since very little change takes place in the recovered audio under varying signal conditions. Lack of synchronous local oscillator operation may present problems with tone control or signaling equipment unless it will tolerate frequency errors of 50-100 Hz. Errors this great are unusual but could develop in unattended equipment over a long period of time. (In operating a CF-1 system over a two-hop radio circuit we found that the drift between terminals on the worst channel was on the order of about 12 Hz per year.)

If it is necessary to transmit accurate tones on CF-1 equipment, channel 1 may be used since there is no conversion on this channel. Narrow-shift tone channels may also be inserted between 2800 Hz, the upper edge of channel 1, and 3100 Hz, the upper edge of channel 2. Control tones could also be added above 11.8 kHz. Figure 3 shows the channel allocations used with CF-1 carrier systems. Carrier systems cannot transmit on the same channel frequency in both directions over the same pair of wires. The transmission path in the reverse direction must be on a different pair of wires or on

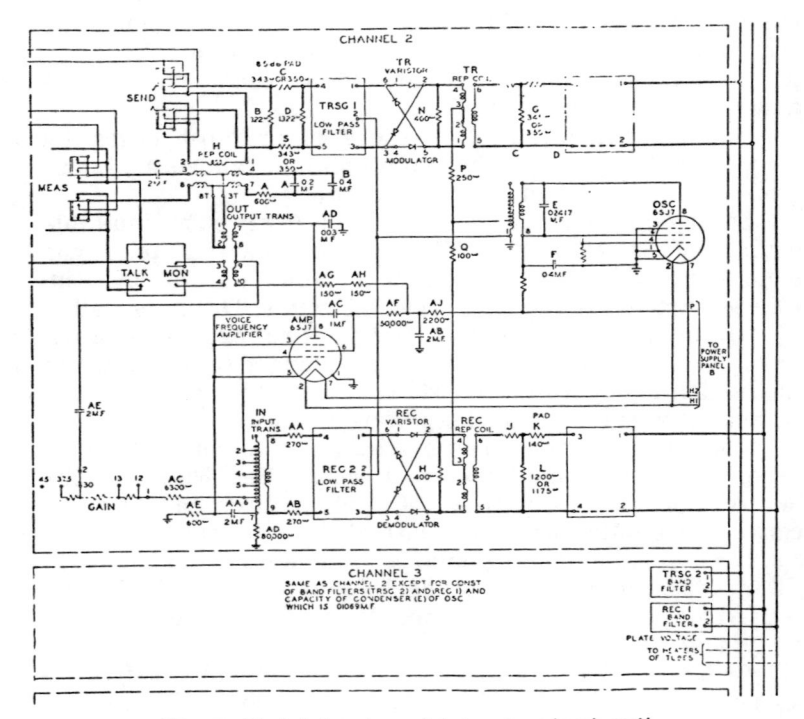

Fig. 2. Modulator-demodulator (modem) unit.

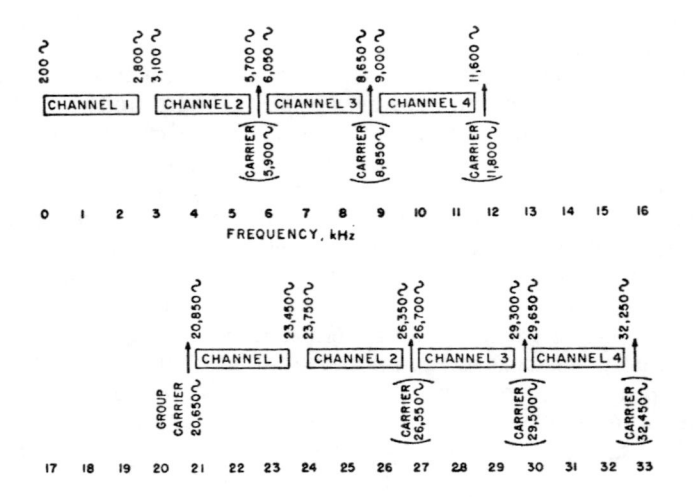

Fig. 3. Multiplex frequency allocation.

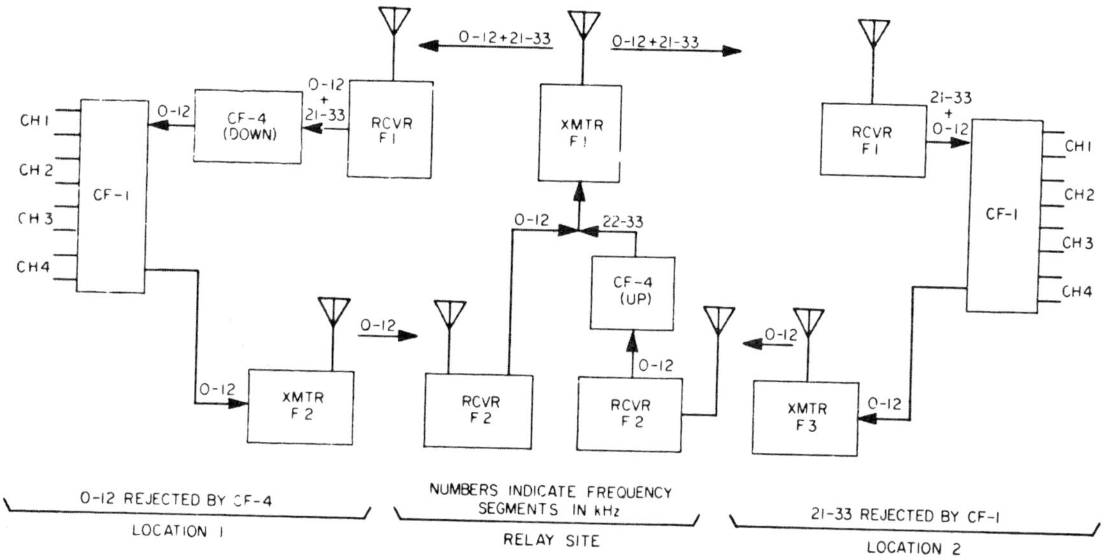

Fig. 4. Deployed multiplex system.

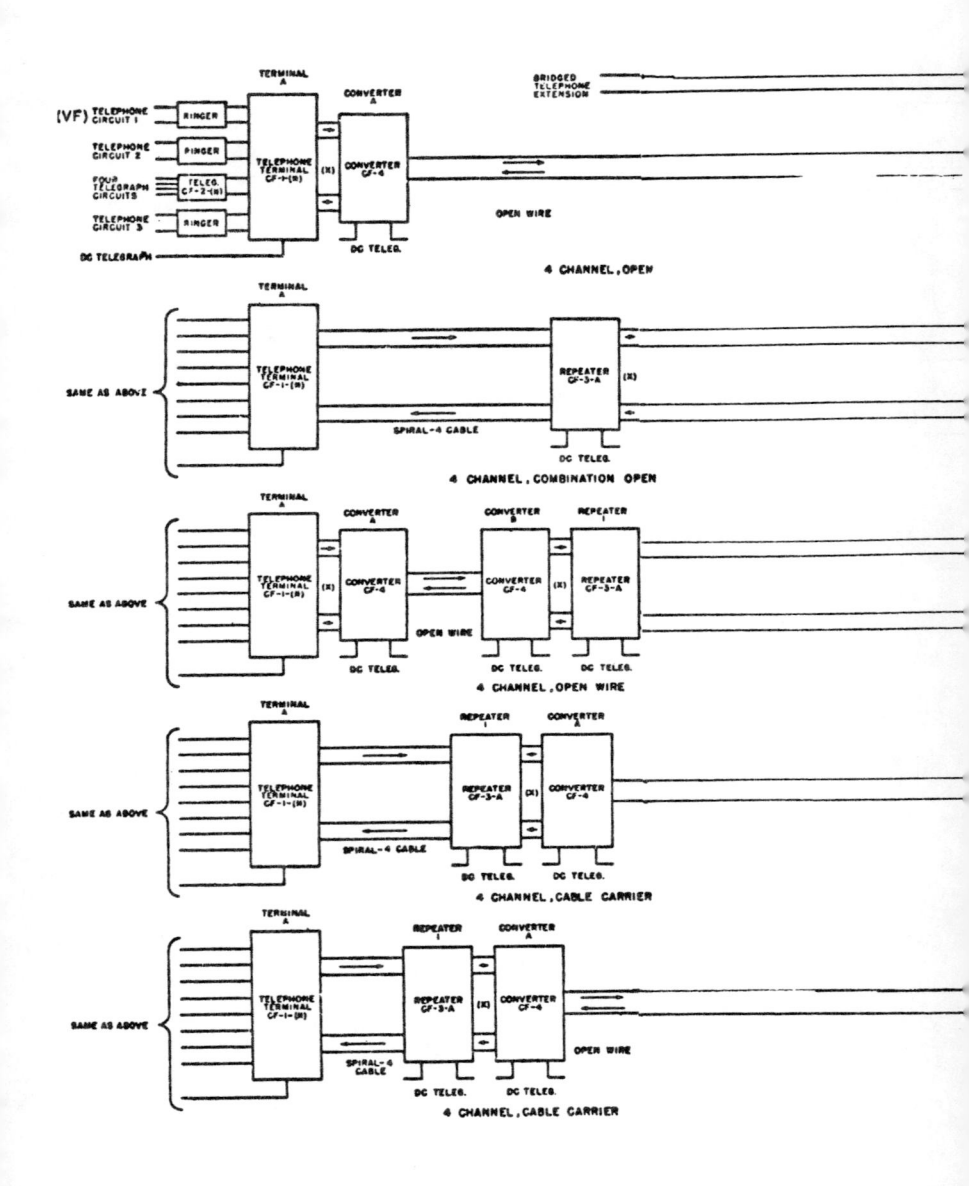

different channel frequencies. The CF-1 was designed to use spiral-four cable transmitting in one direction on one pair and the other direction on the other. The CF-4 converter allows transmissions in both directions over open wireline. The high group allocation between 20 and 33 kHz is used when a CF-4 converter is added to the system. Radio systems are usually

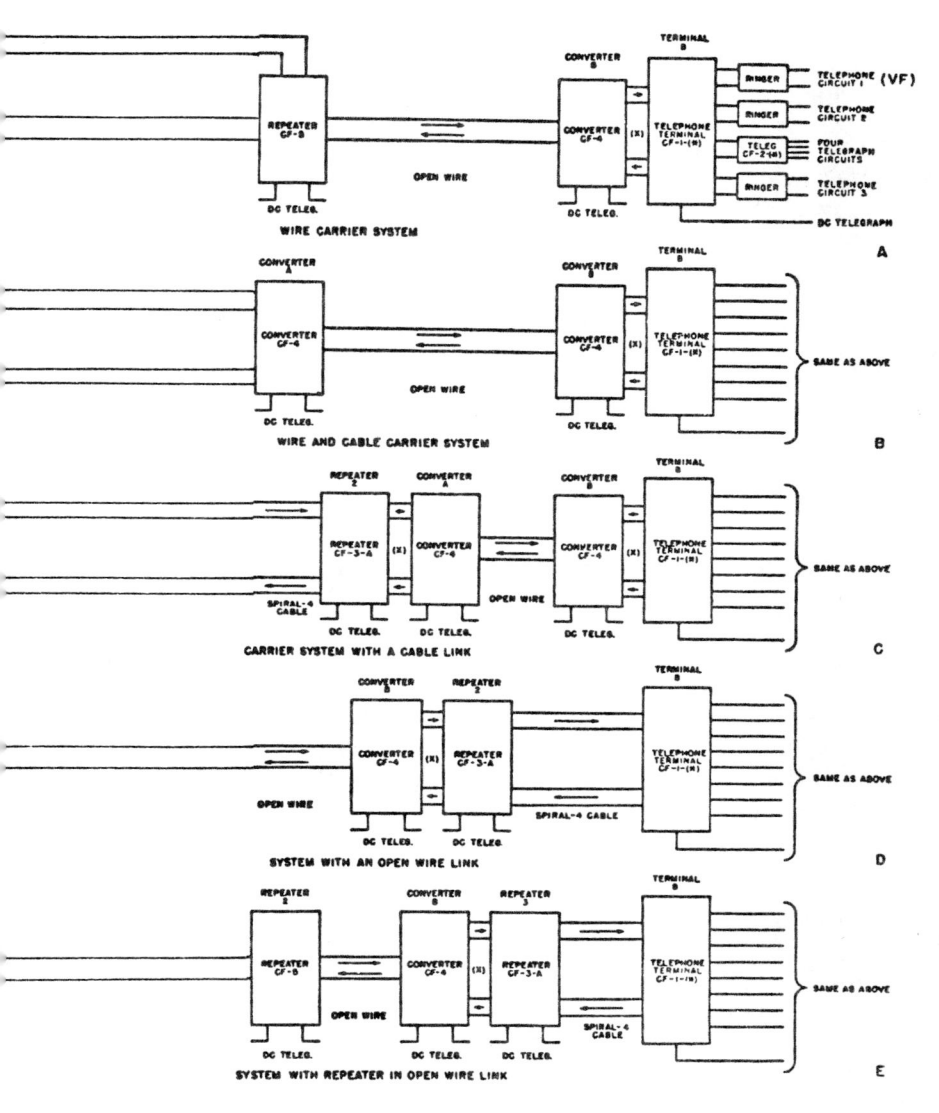

Fig. 5. Open-wire multiplex system using CF-4 unit.

one-way devices in that they can normally operate in only one direction at a time. Transmission in the reverse direction would be on a different radio channel (pair of wires), making a "four-wire" system. When both directions will work at the same time between two terminals the system is known as four-wire, full duplex.

253

Another carrier system that has turned up on the surplus military equipment market is the TCC-3 (also the TCC-7, which is similar but will handle 12 channels). The TCC-3 system has one straightthrough channel and four carrier channels spaced 4 kHz apart. This equipment is about one-fourth the size and weight of the CF-1 and requires more ac power for operation. The TCC-3 is much newer than the CF equipment and is also more expensive (when it can be located).

On the current commercial market are several types of multiplex equipment that serve what is called light traffic routes. Most of the telephone suppliers such as Lenkurt, Lynch, and GE have light-route systems. One of the newest systems is the Cardion 29B multiplex and 22B, transmit-receive device, a combination 960 MHz radio system and light-route carrier terminal. This equipment is designed to meet the new requirements of commercial point-to-point systems that are being forced to move up from the 450 MHz band.

MAKING ONE TRANSMITTER WORK LIKE TWO

Assume for a moment that a relay site is needed between two locations in a radio system but only one transmitter can be operated at the relay site. It is desired to operate a two-way relay through the site in both directions at the same time. One method might work—using a fast switch in the audio and rf circuits to transmit a short time in one direction—then, with the other audio input, a short time in the other direction, etc. The switching rate would have to be about 8 kHz for ordinary voice bandwidth.

A better method would be to use multiplex. The transmission from location A through the relay station to location B would be on the regular voice frequencies (or what would be channel 1 on a CF-1 system) and the transmission from the relay station to location A in the other direction would be on a carrier channel through the same transmitter! The transmitter and antenna system would have to be capable of reaching both locations at the same time.

Going one step further, suppose that a CF-1 system was operating between these two locations under the same conditions. Both receivers at the relay site receive channels 1 through 4 from their respective terminal locations. The channels from location A are fed to the relay transmitter as received without any conversion. The channels from location B are fed to a CF-4 converter operating to convert the four channels up to the high-group allocation. These channels are then added to the other four channels and transmitted on the same radio channel. At the second location, a CF-1 terminal is used to modulate (and demodulate) the system. At location A, the CF-1 modulates four channels and couples them to the transmitter feeding the relay site. The receiver at location A feeds another CF-4 operating as a down converter translating the high-group channels into signals that can be fed to the CF-1 for demodulation. This system is shown in Fig. 4. Figure 5 shows a block diagram of a complete CF-4 open-wire system.

No provision is included in the amateur rules (Part 97) for multiplex operation. Multiplex is allowed in other services on frequencies that are authorized for F3 emission. However, new rules for commercial services are pushing the use of multiplex up into the 900 MHz region and above. The FCC has not been averse to use of multiplex on the UHF amateur bands and has indicated that authorization would be considered if a need for it can be shown. There are several amateur radio remote control systems using multiplex now licensed in the United States.

The Inside Story of the $7 Gem

Back in the early days of crystal making, the crystal manufacturer cut his "rocks" with little more than a protractor to determine the angle of his cuts. Today, with rigid constraints of tolerance, temperature, and size becoming increasingly important, the manufacturer must rely on a vast array of complex measuring instruments and test equipment.

In spite of all this, crystal making is still an art rather than a science. The crystal maker knows that with the right oscillator circuit he can provide a crystal slab with a "grain" angle and finish so precise that frequency tolerances to ± 0.00025 percent can be held without the use of a crystal oven.

Stability

With miniaturization the order of the day, circuit space is at a premium. And on many late-model high-band radios, no provisions have been included for space-consuming crystal ovens. Instead, the radio manufacturers provide "temperature-compensating" circuitry so the burden of stability rests more than ever with the crystal rather than the oscillator. To satisfy these demanding frequency-tolerance requirements, crystal elements must be held to within one-half a minute of the intended angle of cut. This follows through all stages of lapping and polishing, where angles may shift upward by as much as 4 minutes.

Fortunately for those in the crystal-making business, most FM'ers and two-way people, the heaviest crystal users, are already keenly aware of the fact that crystals are not absolute frequency-determining elements. Most of the "off frequency" complaints of other classes of crystal users could be solved if they were as knowledgeable as FM'ers about the effects of an oscillator circuit on a given crystal.

Oscillator Differences

To demonstrate how the frequency of operation is altered when a given crystal is used in various circuits, one crystal manufacturing company actually built prototype oscillator circuits of various configurations—Colpitts, Hartley, Armstrong, and several others—and used one crystal for all of them. The findings: When multiplied to 50 MHz, the oscillator frequency varied by as much as 15 kHz!

The plain truth is: most calibration "error" tolerances required for commercial two-way equipment are in the neighborhood of ±0.002 percent. Thus, even though the GE transmitter would use the same frequency multiplication factor as Motorola, the GE crystal would not "zero" in the Motorola unit.

There are two basic parameters that affect a crystal's frequency of oscillation: crystal drive and oscillator load capacitance. So profoundly does load affect the oscillating frequency that virtually all commercial units are equipped with a means for effectively varying it to allow some degree of crystal "rubbering." But even "rubbering" allows only a finite amount of shift before oscillator performance begins to fall off.

From all this, one can readily see the importance of supplying the crystal maker with sufficient oscillator data when ordering a new crystal. From my own experience in radio, I can stake, unequivocally, that you cannot "oversupply" the crystal manufacturer with information. When ordering crystals, make it a point to include:

- equipment manufacturer and model number
- equipment type and part number
- operating frequency
- crystal frequency
- information about circuit (oven or nonoven use?) (does unit have AFC?)

All is not lost, of course, if you don't have all that information. The supplier can provide the proper crystal by just looking at your oscillator circuit. You can Xerox it or sketch it on a sheet and send it with all the supporting data you have about the unit in which the crystal is to be used.

Natural vs Synthetic Quartz

The piezoelectric material that is the heart of any crystal can be obtained naturally or it can be synthesized and mass-produced. A quiet controversy has been going on for years between crystal "experts" who can't come to a complete agreement on the advantages of one type over the other.

There is a very loud and powerful voice—that of a very large number of crystal producers—making a very convincing case for synthetics. Synthetic material is cheaper, which means a lot more crystals for a lot less money. The material is softer; it can be ground and cut and graded with unbelievable speed. With careful control, synthetics can be made to perform almost exactly like the natural stuff. The crystals—if they are supplied from a quality-conscious manufacturer—exhibit characteristics that are indiscernible from those of a crystal made from natural quartz. The Q of synthetic-quartz crystals is lower than that of its natural equivalent, but this should not be too meaningful if the oscillator circuit is very well designed.

But there is an indefatigable school of hard-headed artisans who just can't quite see it that way. These diehards acknowledge the fact that it takes ten times longer to produce a crystal from good natural quartz, but they say its hardness assures its stability and prevents "aging," a phenomenon they say is characteristic of the softer synthetic quarts. A high Q is important, they say, to guarantee oscillator frequency integrity: the higher the Q, the better the oscillator.

At the risk of making this sound like an ad, I must say that Sentry*was founded on the belief that use of synthetic quartz compromises product quality. And to my knowledge, it is the only company left that hasn't been "won over" by the economic attraction of synthetics. No case for synthetic quartz—however well presented—has ever been able to convince a stubborn artist that natural quartz is a waste of money.

As a matter of fact, Sentry might possibly carry this "nature" thing a bit too far in the quest for perfection: They won't use domestic quartz. With near fanaticism, they make sure that every bit of it coming into the plant is pure Grade A material fresh from Brazil.

Manufacturing

The photos accompanying this article show the birth of a crystal, from the arrival of the raw Brazilian quartz to final inspection and checking of the finished crystal. At Sentry, the raw quartz is inspected soon after arrival for impurities, inclusions and cracks. During this inspection, the optic axis of the rock is determined. The optic axis is the "grain" of the crystal—the reference point from which various axial cuts are based. The three basic angles are on the X, Y, and Z axes; Z is the optic axis. Such factors as expected temperature environment, circuit tolerances, and physical size requirements will determine the axis on which a given crystal will be cut.

---

* Sentry Crystals, 1634 Linwood Blvd., Okla. City, Okla. 73106.

Fig. 1. The first step in a crystal's life is orientation for optic (Z) axis.

Fig. 2. The squares are x-rayed for specific angle.

After inspection, the quartz stones are cemented to a glass plate and subjected to x-ray examination, where the optimum cutting angle is determined. Electronically controlled Swiss saws cut the stone into wafers, which are in turn diced into a variety of square sizes.

Ironically, after they're squared, the wafers are rounded. And here's where tolerances **really** start getting tight. Diameters are held closer than plus or minus 50 millionths of an inch—about the thickness of the fog of your breath on a sheet of glass.

After rounding, the next steps are rough, intermediate, and fine lapping to the approximate frequencies. Some units, such as those for low frequencies, must be beveled. Others, for

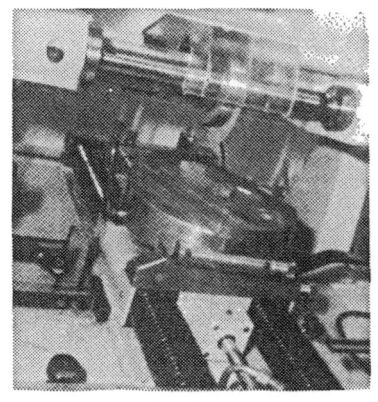

Fig. 3. Crystal hunks are then sliced into wafers.

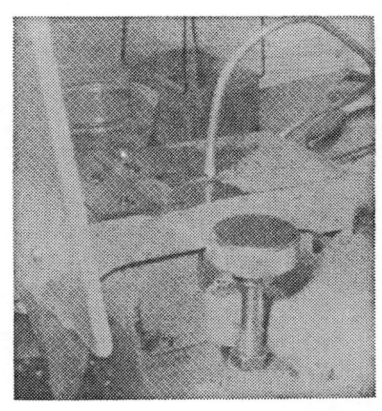

Fig. 4. The wafers are diced into squares.

259

Fig. 5. Following the x-ray, each crystal is loafed and rounded.

Fig. 6. The crystal is then transferred to another department for lapping and calibration.

Fig. 7. Some high-frequency crystals need a high-speed polish.

Fig. 8. The etching process removes all traces of particles and dirt.

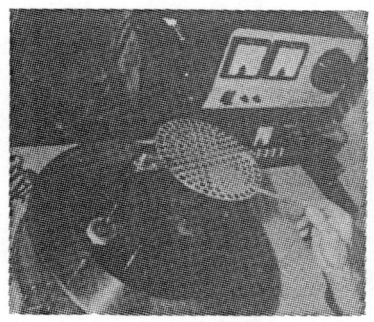

Fig. 9. After all traces of particles and dirt have been removed, the crystal can be keyed with a base plating of gold, silver, or aluminum.

Fig. 10. The crystal is mounted and bonded to a base fixture.

Fig. 11. Final calibration is accomplished with a nickel solution.

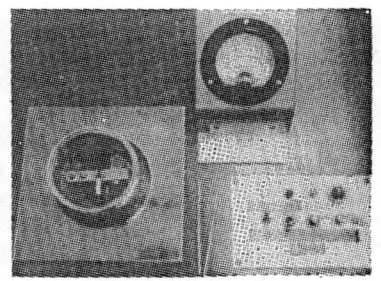

Fig. 12. The can is evacuated and filled with some form of inert gas.

overtone or high frequency use, are polished to an ultratransparent flat finish. After the lapping stages, the quartz blanks are etched, ultrasonically cleaned, and then baked. Next, the crystals are metal-plated under high vacuum using silver, gold, or aluminum. The metal key becomes an electrode on the quartz blank. After plating, the crystal is mounted on the proper base and cemented at both edges of the spring mount using a conductive, thermal-setting cement.

The crystals are cured, and the unit is calibrated to final frequency by plating the electrodes in a nickel solution. After baking the plated crystal, the can is attached, Finally, the entire unit is evacuated, and either sealed off under vacuum or filled with an inert gas such as nitrogen or helium.

People often ask why all Sentry crystals are plated rather than pressure-mounted. The plated crystal has these definite advantages.

- Maximum piezoelectric coupling is attained.
- Possibility of arcing between electrodes and crystals is minimized.
- Frequency change due to shift of relative positions of crystal and electrodes is eliminated.
- Tighter calibration and temperature tolerances can be achieved.

Fig. 13. Crystals for special high-precision use get the Coldweld treatment.

Fig. 14. Before final inspection.

The only real disadvantage is that the plated crystal cannot dissipate as much internal heat as can the pressure-mounted unit. For this reason, plated crystals must operate at lower drive levels than pressure-mounted units of the same cut and frequency.

All Sentry crystals are checked at final inspection for resistance to vibration and shock; absence of moisture, cracks, or inclusions in the base and seal; resistance / activity; unwanted modes; and excessive pin-to-pin capacitance. Crystals are subjected to a number of checks which may include testing under a variety of temperature conditions and in oscillators which are exact electrical duplicates of those in which the crystal is to be used. Final frequency determination is done with an electronic counter to be sure the crystal operates within the specified tolerance.

When you stop to consider all the processing a high-quality crystal must undergo before it reaches your oscillator, the 5- to 7-dollar price tag doesn't seem very steep at all, does it?

Touchtone—How to Use it for FM Control

by Gene Mitchell K3DSM

The purpose of this article is to give FM'ers a basic understanding of Touchtone operation so that they might experiment with the possibilities of using it in conjunction with FM remote control applications. Touchtone is Bell Telephone's system for telephone dialing at a considerably increased speed.

The Touchtone System

Two tones are dispatched to the central telephone office for each digit selected (corresponding to a dialed digit). The lost time of waiting for the 10 pps pulse train after each dialed digit is regained because the system does not require the sequential transmission of a single series of contact breaks.

Many Touchtone dials have been made available to the surplus market, ranging from the standard 10-button dial to the 12- and 16-button military and computer versions. Figure 1 shows the 25A3 10-button dial with seven leads terminating at the connector. These leads may be used as shown to convert the system to two-wire operation in an arrangement similar to a transistor-microphone interconnection scheme. The supply voltage is fed to the dial over the same path as the output of the tones. Figure 2 shows a method by which the Touchtone dial might be used with an FM transmitter.

In the sketch of 2a, the transmitter mike amplifier is shown along with the input connector. Operating voltage is 9-12V across the dial. If you have Touchtone telephone service, a phone patch might be used to make use of Touchtone rather than obtaining a separate dial or telephone. Figure 2b shows the matrix pattern of the tones, including the fourth column (HG4, 1633 hertz), used on 16-button dial units.

The block diagram of the tone receiver shown in Fig. 3 is one used by the phone companies. Since FM amateur application does not require such tolerances as Bell or the other big names, it may be possible to successfully simplify the tone receiver. The entire unit shown—up to the decoder—is one of three basic designs used. The decoder (Fig. 4) is my own

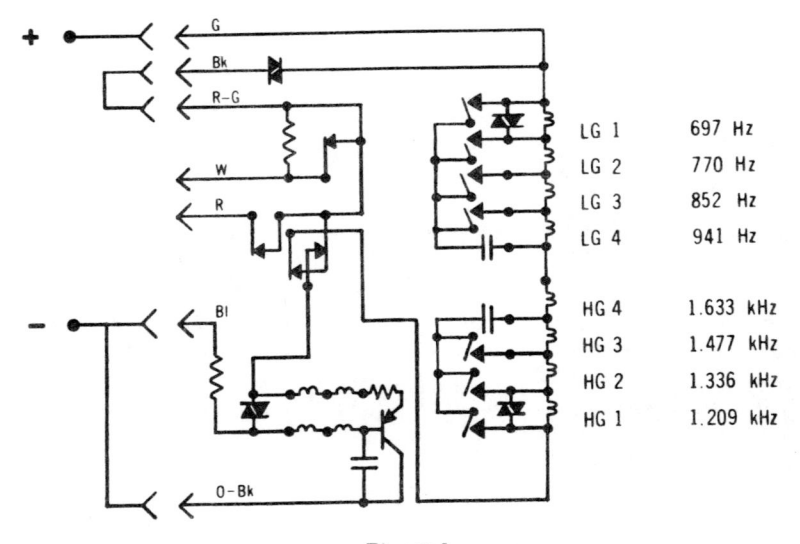

Figure 1

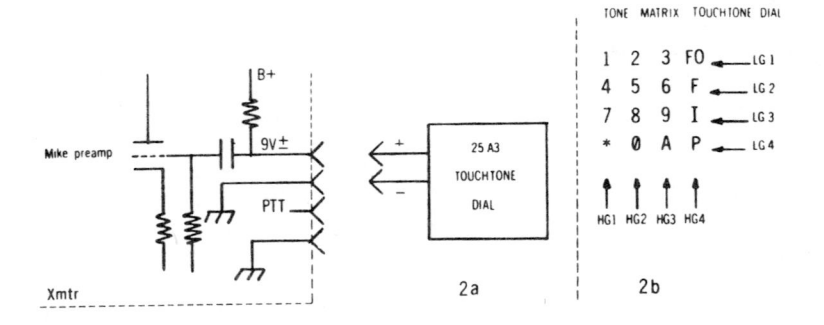

LG 1	697 Hz
LG 2	770 Hz
LG 3	852 Hz
LG 4	941 Hz
HG 4	1.633 kHz
HG 3	1.477 kHz
HG 2	1.336 kHz
HG 1	1.209 kHz

TONE MATRIX TOUCHTONE DIAL

1	2	3	FO	← LG 1
4	5	6	F	← LG 2
7	8	9	I	← LG 3
*	0	A	P	← LG 4

HG1 HG2 HG3 HG4

Figure 2

DECODER BLOCK DIAGRAM

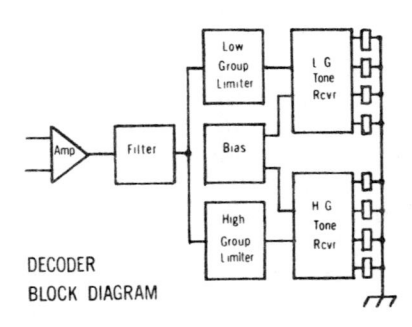

Figure 3

design. Since it may require only a few digits for individual applications, it is possible to eliminate many of the tone frequencies used to simplify construction. Some of the parts used (e.g., transistors and the toroids and capacitors for tone detection circuitry) have no part numbers; this makes some of the values almost impossible to acquire. However, with some experimentation and variation to the circuit, it should be possible to determine what is necessary to make it work. The relays (Y0-Y9 and Z1-Z4) are Western Electric 295A reed relays with a coil resistance of 525 ohms. These relays have 6-pole-single-throw output.

It is possible—and practicable—to use other relays and provide the necessary 4-pole contacts needed on the high group and single-pole contacts needed on the low group. The decoder shown is set up for all eight tones (16-button dial).

Symbols used in the decoder for relay coils and contacts are standard in the telephone industry:

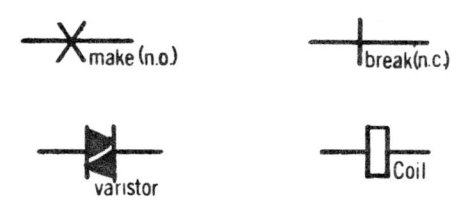

—✕— make (n.o.)   —┼— break (n.c.)

—▶|— varistor   —⊐⊏— Coil

The sample code shown in the decoder (593) may be changed to fit applications necessary by rearranging the

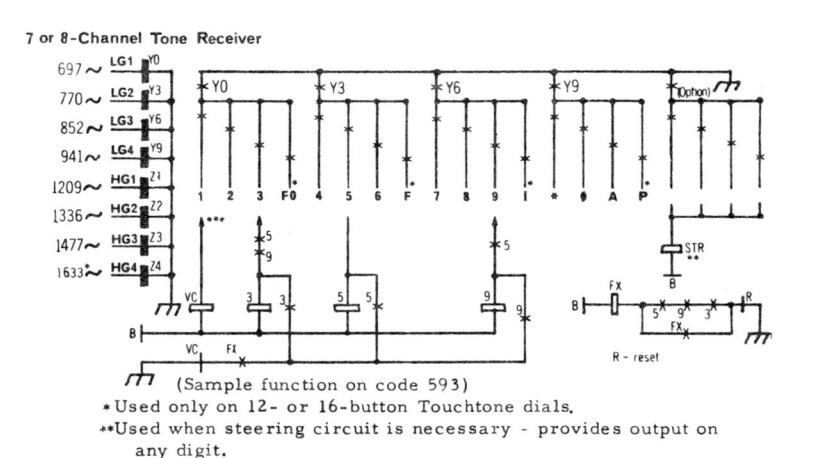

7 or 8-Channel Tone Receiver

(Sample function on code 593)

*Used only on 12- or 16-button Touchtone dials.
**Used when steering circuit is necessary - provides output on any digit.
***Unused digits are strapped together and to VC relay.

Figure 4

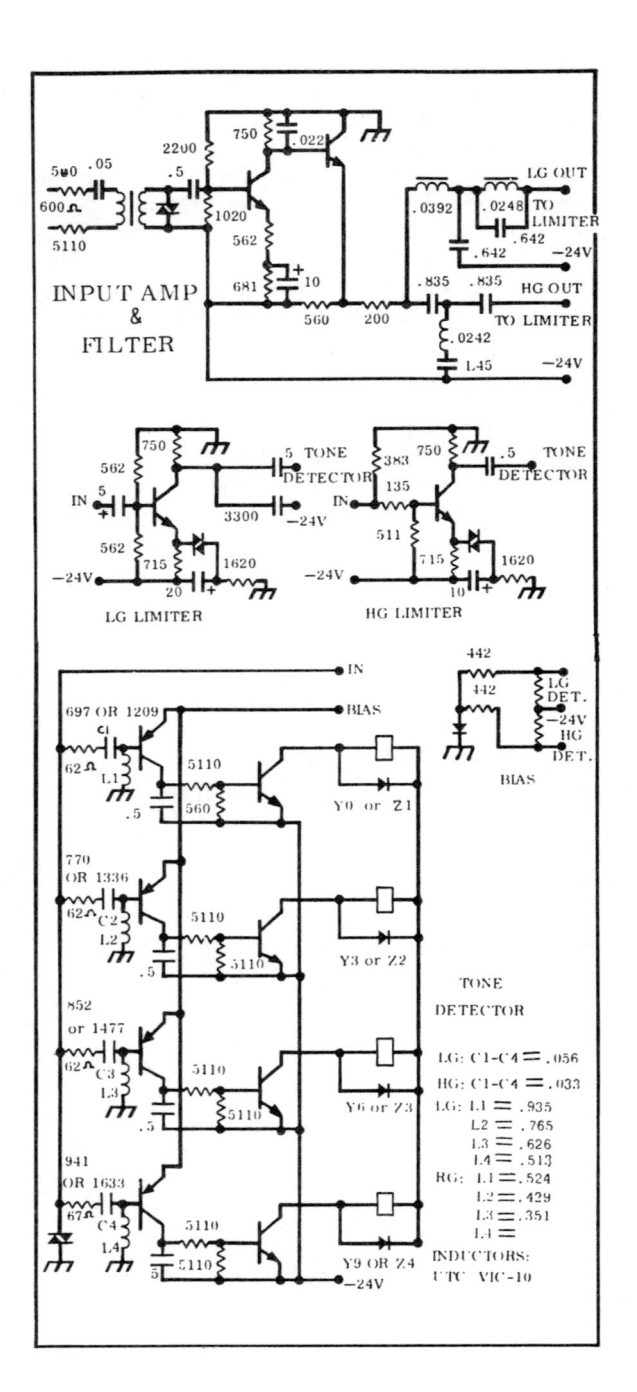

Figure 5

counting relays. Relay 5 operates on proper tones, giving a partial path for the final function and the last digit (9). Relay 9 must operate next to give more of the path to the final function and the next digit (3).

After relay 3 operates, the final function relay operates and holds until released by another function or digit (such as 1).

If the tone receiver contains all of the tones or at least a 10-digit output, the unused digits can be strapped together to operate the VC relay which will drop any of the counting relays should someone start random dialing to trip the function.

With the arrangement shown, no digits can repeat. If this is desired, a steering circuit must be added to prevent the first digit-to-be-repeated from operating the next relay of that number. A code such as 353 can be used designating the second 3 as 3' (3 prime). It would be wired to operate only after the 3 and 5 relays are up.

By application, you might want to wire a timer to the tone receiver so that it only looks for the tone signals in the first 10 seconds after the receiver's carrier-operated relay has been operated.

For the information of prospective Touchtone builders, circuits are provided herein for the various functions shown in the block diagram. These schematics are shown in Fig. 5.

Dials & Switches
and
Things Like That

Although it is true there is but little to say about a simple telephone dial, it is an extremely important part of most remote control systems, and virtually indispensable with remotely operated telephones. (Remember, it may be a telephone dial to you, but to the repeater owner, it's a digital formatter.) So, what is to be said **should** be said, and it seems appropriate to say it first.

A dial is a circuit interrupter (and NOT a contactor as many believe). When the finger plate is released after being turned from its normal position, it interrupts the line circuit in quick succession a number of times corresponding to the digit dialed.

Thus, if a dial is to be used in a circuit that requires a series of "makes" for pulsing, the dial will have to be used to drive a normally closed relay by holding it "off-normal" until pulsed.

Dials typically have several sets of contacts that make while the dial is in motion and break as it comes to rest. These contacts are usually employed for keying the push-to-talk of the transmitter in remote applications. This function is hardly important enough to be mentioned, except for the fact that wiring of the wrong contacts could result in automatic dropping of the final pulse each time the dial is used. Just remember to use the set of contacts that stay in contact until after the pulse sequence has been transmitted.

In many—perhaps most—amateur radio control schemes, the dial at the control point pulses a tone encoder so that a series of beeps is transmitted. The beeps are decoded at the remotely situated receiver to yield a series of relay closures. The relay closures are typically used to drive a rotary stepper switch of one kind or another.

And this brings us to the next point of discussion. The rotary stepper switch is the basis of much of today's automatic telephone switching. The switch is of the ratchet type, consisting essentially of one or more wiping springs fixed on a shaft which is moved by a pawl-and-ratchet mechanism (Fig. 1). This mechanism is actuated by an electromagnet, which

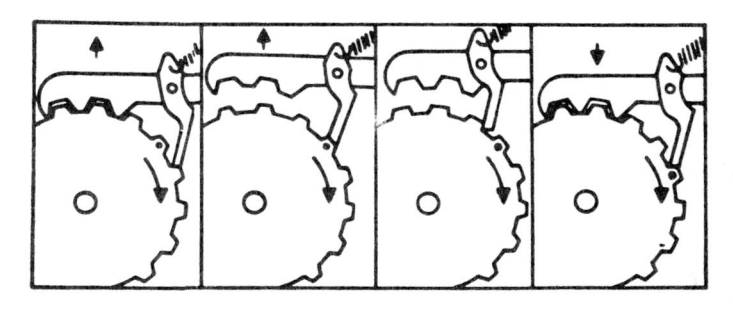

Figure 1

responds to momentary surges of current from the pulsed decoder. At each pulse, the pawl engages the ratchet, moving the wipers one step forward over a bank of contacts. Figure 2 is a diagram of a spring-driven rotary stepper switch. There are two types of driving mechanisms associated with rotary steppers: indirect (spring-driven) and direct.

Spring-Driven Stepper. Operation of the stepping magnet moves the driving pawl out of the ratchet and drops it over the succeeding tooth, but does not move the wiper assembly;

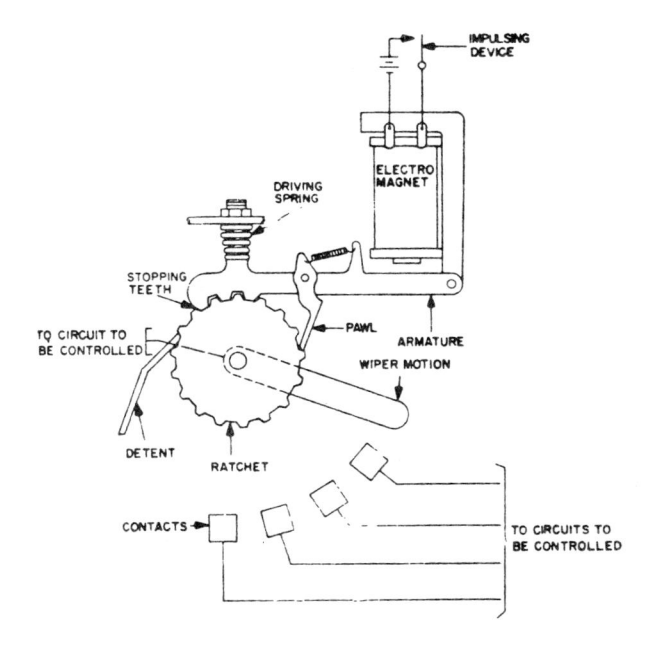

Figure 2

when the magnet is deenergized, the wiper is driven forward by a spring. The switching operation is not complete until the magnet circuit is opened.

Direct-Driven Stepper. Operation of the stepping magnet moves the pawl into the ratchet and moves the wiper assembly; a detent holds the wipers in place when the stepping magnet is deenergized and the driving pawl is returned to normal by a spring. The switching operation is complete the moment the magnet operates.

Probably the most universally functional of the steppers is the Strowger (after inventor A. B. Strowger). In commercial telephone service, the Strowger type of switching device (Fig. 3) is the principal mechanism used in establishing automatic connections. This switch selects one out of 100 possible contacts. A contact bank consists of ten levels with ten contact points at each level. The wiper moves in two axes by being driven vertically on the first pulsing sequence, then horizontally on the next. The first digital sequence drives the

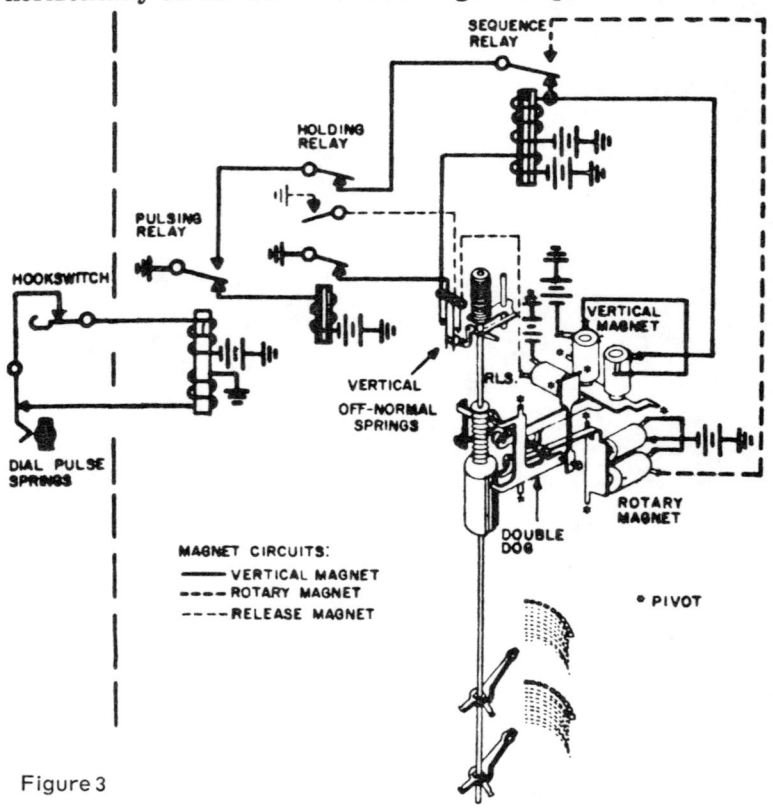

Figure 3

wiper to the proper deck; the second selects the proper contact of that deck.

How does the Strowger know when to step up and when to step across? The partial schematic of Fig. 3 shows the typical use of fast and slow relays to channel dial pulses to the proper stepping magnet. In telephone use, when the caller lifts his handset, the hookswitch closes the line circuit and operates the pulsing relay. The pulsing relay operates the holding relay, which in turn operates the sequence relay. In amateur use, the hookswitch function is generally accomplished by the contacts of a carrier-operated relay.

Index

A

Add-on amplifier	33
Adjust, frequency	126
AFC	190
AFC-type oscillator	176
Alignment, receiver	188
Alternate-channel	231
AM	118
Amateur	
—FM	263
—UHF	39
Amplifier	
—add-on	33
—circuit	114
—efficiency	35
—mike	263
—power	205
Antenna	203
—directional	231
—logarithmic beam	11
—range	25
—sewerpipe	22
—stage	186
AREC	182
Arrays, commercial	11
Audio circuitry	129
Autopatch	43
"Auxiliary" receiver	122

B

Band, two-meter	124
Base station, FM	184
Batteries, Ni-Cad	58
Battery	
—D-size	199
—lead-acid	196
—nickel-cadmium	194
Beam, frequency independent	10
Bias voltage	109

Block diagram	263
Board, oscillator	124
Bogus ovens	193
Brackets, trunnion	202
Bumper mount	201

C

Cadmium	
—electrode	197
—oxide	197
Calibration oscillator	77
Capacitor	128
—trimmer	127
Cascode, FET	169
Cavity loop	206
Cell	194
Celsius	193
Channel	
—crystal accuracy	79
—frequencies	252
—input	229
Charging	198
Circles, RTTY	47
Circuit	
—amplifier	114
—interrupter	268
—transistor	37
Circuitry, squelch	214
Clever estimating	209
Coax-section collinear	8
Cockpit	202
Coding, tone	230
Coil, relay	265
Collinear	
—coaxial-section	8
—two-meter	8
Commercial arrays	11
Communicator, FM	133
Compensation, reactance	10
Compressor-amplifier	44
Com-Prod station master	13
Construction	65

Contact, 4-pole	265
Contacts	268
Control	
—crystal	123
—head	202
Conversion	
—power supply, 6 to 12V	179
—6 to 12V	177
Converter, touch-to-secode	267
COR tube	37
Cord, mike	203
Crystal	126
—accuracy, channel	79
—control	14
—oscillator	50
—ovens	192
—two-meter	126
Crystals, sentry	200, 262
CTS decoder	82

D

Decode, solid-state	37
Decoder	39, 42
—CTS	82
"Depth of charge"	194
Detectors, phase	90
Device	
—low-current	37
—timing	221
Diagram, block	263
Dial, 10-button	263
Dials	268
Digital	
—encoder	46
—identifier unit	62
Diode matrix	63
Diodes	56
Dipoles, nonresonant	11
Direct-driven stepper	270
Directional antennas	231
Division multiplexing, frequency	246
DPDT	145
D-size battery	199
Dual-tone	46

E

Electrode, nickel	197
Electronic voltmeter	192
Encoder	39, 42
—digital	46

F

FET cascode	169
Filament transformer	45
Filter, whine	45
Filters, off-frequency	190
FM	147, 200
—amateur	263
—base station	184
—communicator	133
—handie talkie	28
—receiver	120, 188
—relay	213
—service center	204
Frequencies, channel	252
Frequency	
—adjust	126
—division multiplexing	246
—independent beam	10
—meter	76
—stability	39
—standard	55
—synthesis	85
Functions, timer	221

G

Gassing	197
GE	
—4ER6	143
—master	157
—Pre-prog	157
Generator, signal	127
Grid-dipping	28
Gutter clamps	201

H

Half-wave rectifier	199
Handie-talkie	167, 182
—FM	231
Head, control	202
High-band	122
—oscillator	122
Hybrid	
—coupling	234
—patches	44

I

Identifier unit, digital	62

273

Ignition	
—noise	205
—tuneup	203
Inductance, transmitter	185
Input channel	229

J

Jack, meter	146

L

Land-mobile radio station	200
Lead	
—acid battery	196
—polyvinyl-jacketed	127
Leakthrough	34
Logarithmic beam antenna	11
Loop	
—cavity	206
—phase-locked	89
Low	
—band	122
—band receivers	123
—current device	38
—i-f	162
—power repeater	38

M

Makes	268
Manufacturing	258
MASTR	216
Message mate	120
Meter	
—frequency	76
—jack	146
MHz	40
—quarter-watt	182
—signal	120
Microvolt	123
Mike	
—amplifier	263
—cord	203
Mobile	200
—operators	26
Mod II unit	42
Modulator	186
Module, oscillator	124
MOTRAC	216
Multiplex	245

Multiplex versus transmitters	246

N

NAND gate	61
Narrowband	157, 212
Narrowbanding	
—receiver	160
—transmitter	159
Natural vs synthetic quartz	257
Ni-Cad	194
—battery	58, 196
Nickel-cadmium battery	194
Nickel electrode	197
Ninic	127
—pocket communication receivers	120
Noise, ignition	205
Nonlinear waveform	107
Nonresonant dipoles	11

O

"Odd ball" unit	193
Off-frequency filters	190
Ohmmeter	192
Oscillator	186
—AFC-type	176
—board	124
—calibration	77
—crystal	50
—differences	256
—high-band	122
—module	124
—"ring"	43
Oven	
—bogus	193
—crystal	192
—pins	192
Overcharging	196
Overdischarging	194

P

Patches, hybrid	44
Peaker-tweaker	74
Peaking, receiver	184
Phase detectors	90
Phase-locked loop	89
Phone system	40
Pins, oven	192
Plate 6DL4	164

Pocket
—receiver 120
—tucawayable 120
Polyethylene rope 19
Polyvinyl-jacketed
  lead 127
Power
—amplifier 205
—quadrupler 17
Pre-prog receiver 164
Prodelin Omni-6 13

Q

Q reduction 10
Quarter-watt, MHz 181
Quarter-wave 262

R

Radio frequency shielding 99
Range, antenna 25
Reactance compensation 10
Realignment 190
Receiver 151, 186
—alignment 188
—"auxiliary" 122
—FM 120
—narrowbanding 160
—ninic pocket communi-
  cation 120
—peaking 184
—pocket 120
—pre-prog 164
—squelched 203
Rectifier
—half-wave 199
—tube-type 37
Reduction Q 10
"Refrigerator-light" 201
Relay
—coil 264
—FM 213
Remote telephones 234
Repeater 226
—low-power 38
—operations 69
—UHF 40
—"walkie-talkie" 38
Requirements, system 40
RF stage 186
"Ring" oscillator 43
Rope, polyethylene 19
Rotary wheel 47
RTTY circles 47

S

Schmitt trigger 49
Secode system 46
Selectivity 188
Semigrid 25
Sentry crystals 200, 262
Service center, FM 204
Sewerpipe antenna 22
Shielding 113
—radio frequency 99
Signal
—generator 127
—MHz 120
—transmitted 206
Single-frequency
  encoders 47
Slugs 126
S-meter 184
Socket, tube 164
Solid-state
—decoder 37
—timers 224
"Speaker" 46
Spring-driven stepper 269
Squelch 129, 133
—circuitry 214
—zener stabilized 138
Squelched receiver 203
Stability 256
—frequency 39
Stage
—antenna 186
—RF 186
Standard, frequency 55
Stepping magnet 269
Strowger 270
Switches 268
Synthesis, frequency 85
Synthesizer 95
System
—phone 40
—requirements 40
—secode 46
—touchtone 263

T

TD2 44
TD3 44
10-button dial 263
Terminals 184
Timer functions 221
Timers, solid-state 224

275

Timing devices	221
Tone	
—coding	230
—unit	41
Touchtone	232, 263
—to-secode converter	271
—system	263
TPL	216
Transformers, filament	94
Transistor	
—circuit	37
—microphone	263
Transmitted signal	205
Transmitter	256
—conversion	177, 146, 184, 213
—inductance	185
—narrowbanding	159
Trigger, Schmitt	49
Trimmer capacitor	127
Trunk-groove	201
Trunnion brackets	202
T-supply	45
Tube	
—COR	37
—socket	164
—type rectifiers	37
Tuned slugs	34
Tune up	126, 150
Tuneup, ignition	203
Tuning, varactor	243
"Tweak" method	188
Two-meter	
—band	124
—collinear	8
—crystal	126

U

UHF	212
—amateur	39
—repeater	40
Unit	
—Mod II	42
—tone	41

V

Vacuum-tube	37
Varactor	239
—tuning	243
Voltage, bias	109
Voltmeter, electronic	192

W

Walkie-talkie	120
—repeater	38
Waveform, nonlinear	107
Wheel, rotary	47
Whine filter	45
Wide banding	136
W2CVT	230

Y

Y0-Y9	265

Z

Zener-stabilized squelch	138
Z1-Z4	265